Improving Water Policy and Governance

Old forms of governance in both public and private sectors are becoming increasingly irrelevant because of rapidly changing conditions. Because of these changes, both governance processes and the scope of the institutions through which power is exercised throughout society may have to undergo a radical break with the past and prevailing models of governance. Water sector is an integral part of the global system. Consequently, its governance processes and the institutions responsible for its management must change as well in order to cope with the current challenges and potential future changes. Because of these current and future changes, water governance may have to change more during the next 20 years compared to the past 2000 years, if societal expectations are to be successfully met. All these changes will make water governance more complex than ever before witnessed in human history. Improving water governance will require good and objective analyses of case studies from different parts of the world as to what has worked, why and the enabling environments under which good governance has been possible.

The present volume analyses case studies of good water governance from different parts of the world, and for different water use sectors. It concludes with an analysis of the critical issues that should be considered for water governance and a priority research agenda for improving water governance in the future.

This book was published as a special issue of the *International Journal of Water Resources Development*.

Cecilia Tortajada is Scientific Director at the International Centre for Water and Environment, Zaragoza; and Visiting Professor at the Lee Kuan Yew School of Public Policy, Singapore.

Asit K. Biswas is the Founder and the President of the Third World Centre for Water Management, Mexico; and Distinguished Visiting Professor of the Lee Kuan Yew School of Public Policy, Singapore.

Improving Water Policy and Governance

Improving Water Policy and Governance

Edited by
Cecilia Tortajada and Asit K. Biswas

Lee Kuan Yew School of Public Policy
National University of Singapore
Institute of Water Policy

Singapore International Water Week

Third World Centre for Water Management

Routledge
Taylor & Francis Group
LONDON AND NEW YORK

First published 2011
by Routledge
4 Park Square, Milton Park, Abingdon, Oxon OX14 4RN

Simultaneously published in the USA and Canada
by Routledge
605 Third Avenue, New York, NY 10017

Routledge is an imprint of the Taylor & Francis Group, an informa business

This book is a reproduction of the *International Journal of Water Resources Development*, vol. 26, issue 2. The Publisher requests to those authors who may be citing this book to state, also, the bibliographical details of the special issue on which the book was based

Typeset in Times New Roman by Taylor & Francis

British Library Cataloguing in Publication Data
A catalogue record for this book is available from the British Library

ISBN13: 978-0-415-60628-8 (hbk)

Contents

Future Water Governance: Problems and Perspectives

ASIT K. BISWAS & CECILIA TORTAJADA
Third World Centre for Water Management, Atizapan, Mexico; Institute of Water Policy, Lee Kuan Yew School of Public Policy, Singapore

ABSTRACT *One development can be predicted with complete certainty; the world in 2030 will be significantly different from what it is in 2010. Water governance, which is a broad concept, must also adopt to these changes. While there are no usable indicators for water governance that exist at present, some general indicators for governance of individual countries are available. These are of limited value for the water profession. It is argued that at least 10 to 12 good, independent and objective case studies of good water governance would be very useful to learn what were the enabling environment and critical factors that contributed to their success and could allow others to significantly improve their current practices and processes.*

Where is the life we have lost in living?
Where is the wisdom we have lost in knowledge?
Where is the knowledge we have lost in information?

(T. S. Eliot, Nobel Laureate, 1945)

Introduction

Over the past thee millennia, overall governance has progressively shifted from kings and chieftains to a broader coalition of elected representatives, bureaucrats and interest groups representing different sectors of civil society. There is now every indication that the old forms of governance in both the public and private sectors are becoming increasingly irrelevant because of rapidly changing conditions at the global, national and sub-national levels. These changes are due to a variety of factors which include, but are not necessarily limited to, deeper and accelerating global integration, increasing free trade, higher levels of education, rapid scientific and technological developments in nearly all fields, revolutions in information and communication technologies, institutional innovations, growing demographic diversity within countries and between countries, incessant pressures exerted by economic, social and political dynamism, changing societal perceptions, institutional values and structures, and increasing but perceptible demand by the governed of yesterday to become the governors of tomorrow. All of these changes, and more accelerated developments that are likely to be witnessed during the next two decades, mean that there is an urgent need to review the current processes and practices of governance which are becoming increasingly inadequate and ineffective with time. It is highly likely that many of the past (and present) methods of governance, which depended to a significant extent on command, control and obedience, will become less and less applicable and relevant in the future (Michalski *et al.*, 2001). These developments are likely to ensure that in the coming years both governance processes and the scope of the institutions through which power is exercised throughout society may have to undergo a radical break with the past and its prevailing models.

The water sector is an integral part of the global system. Its governance will not be immune to the rapid changes that have occurred, and will continue to occur, both within the governance of the sector and also of other sectors which are likely to affect the water sector, either directly and indirectly. There have to be radical changes in the governance processes and the institutions responsible for water to cope with the immediate challenges, potential future changes and uncertainties both from within the sector and around the sector.

Because of the changes that are likely to take place, water governance has to change more during the next 20 years than it has in past 2000 years if societal needs for water-related activities, including environmental requirements, are to be met successfully in a timely, equitable and cost-effective manner. While many of these changes will come from within the water sector, many more will come from outside the sector which will directly affect this sector and on which the water institutions and professionals, unlike in the past, will have only limited or even no control. They will have to react to these changes and thus these changes have to be correctly anticipated and managed.

All these likely developments will make water governance more complex than ever before witnessed in human history. The water profession will have to seriously consider 'business unusual' approaches and solutions. Yesterday's approaches and solutions, based on past and current experiences, and dependence on incremental progress to solve the global, regional, national and sub-national water problems, will most likely fall far short of what will be required to meet the turbulent times due to the increasing economic, hydrologic and technological uncertainties of the future, and also the accelerating social and political expectations of civil society for a better standard of living and quality of life in all parts of the world.

Water-related Issues

During the past decade, water-related issues have attracted considerably more interest from the media, the general public and political leaders compared to what they had received during the 1980s and 1990s (Tortajada, 2008). This increasing interest could, for the most part, be considered beneficial since it has steadily moved water higher up international and national political agendas. However, whether this move of water higher up the political agenda has had commensurate positive impacts in improving its governance practices and processes is difficult to say. Developments in the water sector probably would have been somewhat similar with or without water moving higher up the political agenda. At best, the improvements have been somewhat marginal because of this move.

IMPROVING WATER POLICY AND GOVERNANCE

Regrettably, a significant part of the discussions in the media in recent years has been on the incorrect water-related issues. For example, much has been written on how the world will face a water crisis of unprecedented proportion in the coming years because of the physical scarcity of this resource, and there will be wars between nations because of acute scarcities of water. Publications on water crises and water wars have become a growth industry! During the past few years, one can easily identify a dozen books in the English language alone on "water crises." The number of articles on the water crisis in the media and scientific periodicals during the past decade would easily run into the thousands in the English language alone. The media also have shown an unhealthy obsession with water wars. Many professionals and political leaders have unnecessarily fuelled this flame because of faulty reasoning, poor knowledge of the water issues, and use of erroneous data. For example, Kofi Annan, the former Secretary General of the United Nations said "Fierce competition for freshwater may well become a source of conflict and wars in the future" (Annan, 2001). In 1988, the then Egyptian Foreign Minister, Boutros Boutros-Ghali, who subsequently became the Secretary General of the United Nations, said, "The next war in the Middle East will not be about politics but over water." Much later, during a 2006 BBC interview, when asked if he was still worried that there could be a war over water, his response was "Yes certainly.... Water will be during this century more important than oil" (Ghali, 2003). On sober reflection, and based on the reliable data and analyses that are now available, such statements are neither correct nor realistic. At best, they are simplistic generalizations of complex issues.

Such erroneous statements by leading public figures, several major international institutions, many water professionals, and most media pundits are based on a simple misunderstanding. Unlike oil, gas and coal which are non-renewable resources and which once used can no longer be reused, water is a renewable resource, which can be used and then reused several times with good governance. Furthermore, in contrast to fossil fuels whose efficiency of uses has undergone remarkable improvements since the 1970s because of increasing demand and higher prices, water continues to remain heavily subsidized in most countries. Consequently, water governance practices continue to be poor to very poor. During the past 35 years, the efficiency of energy usage in most countries of the world has steadily improved. Unlike energy, however, the water sector has not made commensurate progress due to misguided policies and governance practices.

While global figures for water use are extremely unreliable, it is commonly accepted that agriculture accounts for nearly 70% of all water use. And yet very few countries, if any, have realistic agricultural water pricing policies which could contribute to efficient water use. Similarly, in most countries of the world, domestic water supply is heavily subsidized. Consequently, water continues to be overused in both the agricultural and domestic sectors.

Based on research conducted at the Third World Centre for Water Management, it can be confidently said that the world is NOT facing a crisis because of the physical scarcity of water. However, the world will most certainly face a water crisis in the future if the past and current poor to very poor governance practices continue to be used in nearly all developing and developed countries. Surprisingly, at present the world has enough knowledge, technology, management capacity and investment funds to solve its water problems (Biswas & Tortajada, 2009a) within the foreseeable future.

Water Governance

As discussed by Tortajada (2010a) in this issue, governance is a broad concept, has no agreed upon definition, and operates at many levels. Different international institutions such as the Organisation for Economic Co-operation and Development (OECD), World Bank, various United Nations (UN) agencies, and the European Union define governance based on their mandates, interests and biases. However, there are some common features for most of these definitions. These include accountability, transparency, and participatory and decentralized decision making. Furthermore, good governance requires proper allocation and management of resources to collective problems, that is, all citizens receive in a timely and efficient manner requisite public goods of appropriate and acceptable quality.

While the issue of governance has since about 1980 received increasing attention in the development literature, it is a comparative newcomer to the water resources sector. One would indeed be hard-pressed to find any serious discussion of water governance prior to 2000. Between 1980 and 2000, the overwhelming paradigms for the water sector were sustainable water management and/or integrated water resources management, neither of which managed to make any long-term impact on the water sector (Biswas & Tortajada, 2005; Biswas *et al.*, 2005).

The overall definition of governance by the various international institutions has limited value to the water sector especially in terms of the implementation of the concept. In 2003, the Global Water Partnership defined water governance as "the range of political, social, economic and administrative systems that are in place to develop and manage water resources, and delivery of water services, at different levels of society" (Rogers & Hall, 2003).

Based on much of the recent discussion of water governance, a cynic may well ask, "If the word 'governance' is replaced with the earlier focus on 'management', would it make any practical and significant difference?" A cynic may further claim that the water profession has got into a muddle by being unable to distinguish between management and governance, or where management ends and governance begins, or if management is subsumed under governance or vice versa.

Regardless, if one reviews the current water literature, one fact clearly stands out, that is, the era of 'sustainable water management' appears to be coming to an end, or may have even ended. So too is the era of 'integrated water resources management' of the past two decades, which is now being gradually but quietly discarded by national water institutions and international organizations because it has not been possible to implement anywhere in the world for macro- or meso-scale water policies or programme of projects, even though this paradigm has been around for more than half a century in one form or another (Biswas, 2004, 2008a, 2008b). These two paradigms are being rapidly replaced by 'water governance'. Only time will tell if this change is simply the old wine that is being recycled into a new bottle, or if it is indeed a new wine which will have major and perceptible impacts in improving how water is managed or governed in most countries of the world.

Governance Indicators

While there have been dozens of governance indicators in the development literature, four have received considerable global attention. These were developed and continue to be regularly updated by International Country Risk Guide (ICRG), Freedom House,

Transparency International and the World Bank. It should be noted that even the four most popular governance indicators that are used at present are not very useful to develop a road map in terms of how to improve local and regional governance, or how best to improve the governance of any specific sector such as water, energy or agriculture. All of the available indicators of national governance suffer from lack of transparency, problems of inter-comparability between different countries, and even in the same country over time, and suffer from selection biases since all of them are based, to a significant extent, on forms of a composite perception index of experts (Arndt & Osman, 2006). The choice of experts could introduce significant biases in the overall governance indicators. In addition, no corresponding indicator specifically for water governance is available at present which remotely receives similar levels of attention from countries or the media such as the following four overall governance indicators.

International Country Risk Guide

The International Country Risk Guide (ICRG) group has provided its indicators to its clients regularly since 1980. It is an analysis of the potential risks which may be useful for international business operations. It is designed to assess financial, economic and political risks in countries and to compare them between countries (ICRG, 2010).

This indicator contains two components: financial and economic risks and political risks. The first part is based on observed statistics such as levels of economic growth and inflation, per capita GDP, relation of external debt to GDP, and budget surplus or deficit as a percentage of GDP. While it is difficult for most countries to estimate these statistics reliably, the fact remains this part of the indicator is based on observed and estimated data. The political risk assessment, in contrast, is a totally subjective view of the ICRG experts on different pre-selected issues such as the government's potential to stay in office and carry out its declared programmes, investment risks such as contract validity and expropriation, levels of corruption, the quality of the bureaucracy, and the presence of independent legal and law enforcement systems.

ICRG provides its clients with monthly assessments for some 140 countries, and also one-year and five-year scenarios on the basis of best and worst case scenarios.

Freedom House

Freedom House assesses the political rights and civil liberties of countries through its in-house experts. Both types of risk are assessed on a scale of 1 to 7 (1 is the best and 7 is the lowest), and then these risks are averaged to decide if a county is free (below 3), partly free (3 to 5), or not free (6 to 7). Political rights consider electoral process, political pluralism and participation, and the functioning of the government. Civil liberties consider freedom of expression and belief, the right to organise, the rule of law, and individual rights.

For some 192 countries and several disputed territories, Freedom House provides only the annual ratings of the two types of risks.

Transparency International

There is no question that in recent years the governance indicators formulated by Transparency International have attracted the most international attention. Transparency

International began as a small non-governmental organization (NGO) in 1993. The prime mover for its establishment was Peter Eigen, a former staff member of the World Bank. It was established when corruption was not an issue that was seriously considered, at least not openly, by international institutions. The scene changed radically when Jim Wolfenson, who became the head of the World Bank in 1995, decided that corruption is a serious issue which affects a country's economic development prospects.

The first corruption perception index (CPI) of Transparency International was published in 1995. The 2009 index (Transparency International, 2010) covers 180 countries and territories which are ranked in a scale of 0 (perceived to be most corrupt) to 10 (perceived to be least corrupt). The index is based on a composite of different surveys of perceptions of resident and non-resident business people, and expert assessments of the perceptions of the extent of corruption in the countries and territories. In 2009, New Zealand received the highest CPI of 9.4 and Somalia the lowest at 1.1. The 2009 CPI was based on 13 independent surveys, but these surveys did not necessarily include all 180 countries. The CPI also provides a confidence range for each country for its index, and Transparency International (2010) is "90% confident that the true score of the country lies" within this range.

It should be noted that the CPI of a specific country from one year to another changes because perception of corruption has changed, which may not mean the actual level of corruption has changed by the same extent. Because the methodologies for different surveys used to construct the CPI may change with time, one should be very cautious when using the CPI for year-to-year comparisons in the actual level of corruption within a specific country.

Country Policy and Institutional Assessment

The World Bank publishes the Country Policy and Institutional Assessment (CPIA), annual numerical scores based on assessments of the quality of a country's policies and institutions. These assessments have been published every June since 2006. The assessment considers key elements that are within a country's control, and are produced by the Bank staff. The rationale behind the annual production of these assessments is the Bank's belief that "good policies and institutions lead, over time, to favourable growth and poverty reduction outcomes, notwithstanding possible yearly fluctuations arising from internal and external factors" (World Bank, 2010).

The CPIA is derived from 16 criteria that are grouped into four clusters which are given equal weight. The clusters and the criteria are shown in Table 1. Each criteria receives a score between 1 to 6. Unlike clusters, the criteria do not have equal weight. The Bank staff assesses the country's actual performance on each criteria and assign a score. These are averaged to yield a cluster score. The average of the four clusters provides the composite country rating.

Appropriateness of Global Governance Indicators to Water Sector

While the above four commonly used indicators give a reasonable idea of overall governance in a specific country, their potential use for managing the water sector is not only difficult but also fraught with danger. Just as there is no solution that is equally applicable to all of the countries of the world, a country's overall governance performance does not necessarily reflect the effectiveness of its water governance.

Table 1. Country Policy and Institutional Assessment (CPIA)

Cluster	Criteria
Economic management structural policies	• macroeconomic management • fiscal policy • debt policy
Structural policies	• trade policies • financial sector policies • business regulatory environment
Social inclusion and equity	• gender equality • equity of public sector resource use • building human resources • social protection and labour • policies and institutions for environmental sustainability
Public sector management and institutions	• property rights and ruled-based governance • quality of budgetary and financial management • efficiency of revenue mobilization • quality of public administration • transparency, accountability and corruption in the public sector

Source: The World Bank

A good example is the performance of the Phnom Penh Water Supply Authority. An independent, objective and comprehensive evaluation of this Authority (Biswas & Tortajada, 2009b, 2010) indicates that its performance during the past 15 years has been stellar. Many of the performance indicators of the Phnom Penh water supply system are better than London, Paris or Los Angeles, let alone for nearly all developing country cities. Yet Cambodia's ranking in terms of CPI for overall governance is dismal. It ranks 158 in the CPI, along with countries like the Central African Republic, Laos and Tajikstan, and is lower even than Zimbabwe which is ranked at 146. Thus, the governance indicators at the national level may give some general indication of the type of water governance in the urban sector of certain cities, but in many cases these indicators may prove to be completely erroneous.

Unfortunately, no water governance indicator exists at present that can be used with some degree of confidence. Accordingly, it would be useful to develop a usable and reasonably reliable set of indicators that can provide some insight into the quality of the governance of a utility or irrigation department. It is highly unlikely that a perfect all-purpose governance indicator can ever be developed for any one type of water use sector. Nonetheless, efforts should be made to develop such indicators which could help water managers to understand how their performance can be judged, help determine how their performances may be improving or deteriorating over time, and also how they may compare with the performance of other water management institutions in the same country.

It may, however, not be possible to develop an all purpose water governance indicator even for one country. This is because governance requirements for different types of water uses are likely to be different. For example, governance of urban water systems requires very different types of institutions, expertise and approaches compared to irrigation water governance (Molden *et al.*, 2010). The structures and functions of the institutions responsible for these two types of water uses are different. These institutions often belong to different levels of government, and the legal and regulatory frameworks under which they operate are often not the same. Thus they cannot have the same type of governance indicators. This only increases the complexity of developing usable and reliable governance indicators for the water sector as a whole.

Improving Water Governance

Improving the water governance of any water use sector is hampered by the unavailability of good, objective, unbiased and independent analyses of good and replicable case studies. In the absence of analytical, comprehensive and usable studies of good water governance, learning from the advances made by other countries and applying that knowledge has been very limited. For example, at present, the water profession is not even aware of which cities and irrigation-related institutions have made remarkable progress in improving governance during the past decade. Nor is it known what were the main factors of the enabling environment that allowed these institutions to improve their water governance significantly within a period of a decade or so.

International institutions and bilateral aid agencies that have shown an interest in improving water governance have mostly been driven by the assumption of so-called 'best' practices, which have seldom been defined reliably and objectively to be of any practical use. For example, in the area of urban water governance, very few if any institutions have defined what in their opinion constitutes 'best' practices. Does this mean a 24-hour supply of water that can be drunk straight from the tap without any danger, real or perceived, to health? Does this include financial self-sufficiency through appropriate water pricing and cost recovery, or low unaccounted for losses (say less than 10%), or transparency and accountability of the utility, or if the poor have access to clean water, or low but appropriate per capita water consumption (say 100–120 litres per capita per day), or absence of corruption, or consideration of a host of other performance indicators? Or, should a utility be judged best on the basis of achieving important societal outputs like contributions to good health or poverty alleviation in preference to technical and economic indicators which relate to the use of resources (say water and/or financial) with greater efficiency? Should they include both water supply and wastewater management? Should such indicators consider all the above factors in some fashion? These are questions that are important, but, most unfortunately, are not being asked, let alone answered.

Even if such indicators could be developed, and we are not sure that these would be possible, there are some case-specific issues for each country that may not be considered, and yet they could be highly germane in terms of showing the quality and level of water governance. For example, in Mexico, the average stay of a major urban water utility manager is about 18 months, and in India, it is around 36 months. Neither in Mexico nor in India is the head likely to be an accomplished technocrat or manager with a proven track record. Under these circumstances, urban water governance in these two countries is unlikely to improve significantly until good and accomplished heads are appointed and given adequate time to understand the specific complexities and constraints of their the urban water systems, formulate appropriate plans and then implement them. Under the current revolving door policy, it is unrealistic to expect the transient heads to achieve the unachievable. Any good head of a water utility who is realistically expected to improve water governance needs to have a minimum tenure of six to eight years, with appropriate checks on the person's performance. In water governance, like most other fields of governance, success breeds confidence and authority, and this, in turn, breeds more success.

When the 'best' practices are considered, these types of issues are seldom considered explicitly or reliably. In addition, very few terms are defined adequately and applied consistently from one so-called 'best' practice to another, thus seriously undermining the overall governance debate.

There are additional major problems with the 'best' practices approach. All of the current so-called 'best' practices are not selected on the basis of detailed and objective analyses of a large number of water governance case studies from different urban centres. For example, international institutions have the tendency to consider a developing world city to have a 'best' practice of urban water governance primarily when they have given the particular authority a loan. This not only introduces a serious selection bias, but also raises questions about the reliability and objectivity of such a label when the institutions that decide what is a best practice have a vested interest in showing that their loans may have resulted in outstanding results. In the absence of independent and reliable assessments of these best practices, it is difficult to have much confidence in such analyses.

In addition to the selection biases from which the so-called 'best' practices are selected, it ignores many cities that may have good urban water governance but are not being considered because they do not have 'godfathers' who could give them the 'best' practice label. It is essential to cast a wider net to find the 'best' practices from which different utilities can learn some lessons.

In addition, the concept of 'best' practice is fundamentally erroneous. Not only is Voltaire's wise statement "the best is the enemy of the good" valid for water governance, but also conceptually and factually there cannot be a 'best' practice that would be 'best' for the entire world. This is because the world is heterogeneous, with different cultures, social norms, climatic patterns, skewed availability of water and financial resources, management capacities, institutional structures and levels of corruption. The systems of governance, legal and regulatory frameworks, decision-making processes, and types and effectiveness of water governance institutions often differ materially from one country to another, and from one city to another even within one country, often in significant ways. Furthermore, developing countries are at different stages of social and economic development, and their capacity for technology use and uptake are also dissimilar. Their water needs and requirements, as well as water availability, are also different. Accordingly, under such diverse conditions, it is simply unrealistic to expect that a 'best' practice of water governance will encompass all countries and cities with very different physical, economic, social, political, cultural and institutional conditions.

In terms of good and implementable water governance practices, it is imperative that each country or city formulates its own specific strategy based on its special conditions, requirements, expectations and capabilities. It would be dangerous for countries and cities to blindly copy any so-called 'best' governance practice promoted by the various international institutions, without adopting them very carefully and appropriately to suit their own specific conditions and requirements. Water governance strategies must not be structured based on the latest international bandwagon without careful and detailed assessment of their relevance, usefulness and replicability potential within the specific context of the country or the city concerned.

What is needed is a determined attempt to identify cases of 'good' water governance (in contrast to the 'best') from different parts of the world which can then be used to learn lessons which can be considered for possible use, after suitable adaptations and modifications, in other countries or cities. What is urgently needed are at least 10 to 12 independent, objective and reliable case studies of good governance that could form a community of good practices from which countries trying to improve water governance can learn. This type of rare studies are simply not available at present. After analysing such a community of good practices, and assessing their potential replicability, countries or cities can pick and choose what aspects of each good practice can be incorporated in their own water governance strategies.

Concluding Remarks

With this overall philosophy in mind, the Institute of Water Policy, Lee Kuan Yew School of Public Policy, Singapore, and the Third World Centre for Water Management, Atizapan, Mexico, convened the first Global Forum on Water Policy and Governance in June 2009 in Singapore, within the overall framework of the Singapore International Water Week. Eminent experts from different parts of the world were carefully selected because of their acknowledged expertise on water governance and then specially invited to share their views and experiences in this First Global Forum. While the speakers were specifically invited, participation to the Forum was open to any one interested in the subject.

An important component of the First Forum was a brainstorming session to identify research priorities in the overall area of water governance which the Institute of Water Policy, the Third World Centre for Water Management and their partners from different parts of the world could undertake. This was a closed session with only 14 invited participants to ensure free, frank and extensive discussions. An analysis of this brainstorming session can be seen in this issue (Tortajada, 2010b).

Papers published in this special issue include some that were presented at this Forum and subsequently modified appropriately by their authors in the light of the discussions. The rest are papers that were very specifically selected in line with the philosophy outlined before. They analyze and discuss different aspects of water governance from different parts of the world, including their successes, shortcomings and constraints.

References

Annan, K. (2001) Address to the 97th Annual Meeting of the Association of American Geographers. New York. Available at: www.aag.org/news/kofi.html

Arndt, C. & Osman, C. (2006) *Uses and Abuses of Governance Indicators* (Paris: OECD).

Biswas, A. K. (2004) Integrated water resources management: a reassessment, *Water International*, 29(2), pp. 248–256.

Biswas, A. K. (2008a) Current directions: integrated water resources development—a second look, *Water International*, 33(3), pp. 274–278.

Biswas, A. K. (2008b) Integrated water resources management: is it working?, *International Journal of Water Resources Development*, 24(1), pp. 5–22.

Biswas, A. K. & Tortajada, C. (2005) *Appraising Sustainable Development: Water Management and Environmental Challenges* (Delhi: Oxford University Press).

Biswas, A. K. & Tortajada, C. (2009a) Water crisis: myth or reality? *Global-is-Asian Newsletter*, October–December pp. 1–3.

Biswas, A. K. & Tortajada, C. (2009b) *Water Supply of Phnom Penh: A Most Remarkable Transformation* (Atizapan, Mexico: Third World Centre for Water Management).

Biswas, A. K. & Tortajada, C. (2010) Water supply of Phnom Penh: an example of good governance, *International Journal of Water Resources Development*, 26(2), pp. 157–172.

Biswas, A. K., Varis, O. & Tortajada, C. (2005) *Integrated Water Resources Management in South and South-East Asia* (Delhi: Oxford University Press).

Ghali, B. B. (2003) Talking Point: Ask Boutros Boutros Ghali. Available at http://news.bbc.co.uk/2/hi/talking_point/2951028.stm.

ICRG (2010) *International Country Risk Guide*, Available at www.countryrisk.com/reviews/archives/000029.html.

Michalski, W., Miller, R. & Stevens, B. (2001) Governance in the 21st century: power in the global knowledge economy and society, in *Governance in the 21st Century*, pp. 7–26 (Paris: OECD Future Studies).

Molden, D., Lautze, J., Shah, T., Bin, D., Giordano, M. & Sanford, L. (2010) Governing to grow enough food without enough water: second best solutions show the way, *International Journal of Water Resources Development*, 26(2), pp. 249–263.

Rogers, P. & Hall, A. W. (2003) *Effective Water Governance,* TEC Background Paper, Stockholm, Global Water Partnership.

Tortajada, C. (2008) Water management for an increasingly complex and interrelated world, in *Hydropower and Dams, World Atlas*, pp. 11–12.

Tortajada, C. (2010a) Water governance: some critical issues, *International Journal of Water Resources Development*, 26(2), pp. 297–307.

Tortajada, C. (2010b) Water governance: a research agenda, *International Journal of Water Resources Development*, 26(2), pp. 309–316.

Transparency International (2010) Surveys and indices. Available at http://www.transparency.org/policy_research/surveys_indices/about.

World Bank Group (2010) Country policy and institutional assessment. Available at http://web.worldbank.org/WBSITE/EXTERNAL/EXTABOUTUS/IDA/0,,contentMDK:20941073~pagePK:51236175~piPK:437394~theSitePK:73154,00.html

[illegible]

[illegible]

[illegible]

[illegible]

[illegible]

[illegible]

[illegible]

[illegible]

[illegible]

[illegible]

[illegible]

[illegible]

[illegible]

[illegible]

[illegible]

[illegible]

Water Governance in the Middle East and North Africa: An Unfinished Agenda[1]

EGLAL RACHED* & DAVID B. BROOKS**

*International Development Research Centre, Cairo, Egypt; **Friends of the Earth Canada, Ottawa, Canada

ABSTRACT *All aspects of the fresh water situation in the Middle East and North Africa are underlain by the scarcity of fresh water in the region compared with the demands for it. However, physical scarcity has been worsened by institutions that may once have been adequate but that are increasingly failing to meet modern needs for water to be extracted in ways that are ecologically sustainable, used in ways that are economically efficient, and distributed in ways that are socially equitable. Despite this discouraging picture, the thesis of this article is that, in response to both internal and external forces, water-related institutions in MENA are slowly changing in ways that seem likely to improve the situation. Of three key goals identified early in the article—greater attention to demand management, wider stakeholder participation, and adoption of pro-poor strategies—modest gains are evident on the first two, but little on the third.*

Introduction

All aspects of fresh water in the Middle East and North Africa (MENA) are underlain by the scarcity of fresh water in the region compared with the demands for it. The World Bank's flagship development report on water shows that nations in MENA already withdraw nearly 80 per cent of total renewable resources (the next closest region is South Asia, which withdraws less than 30 per cent). It also shows that half of MENA nations withdraw more than 100 per cent (World Bank, 2007; Tables A1.2 and A1.6), which means that they are withdrawing water from non-renewable sources.

Some regions of the world are drier than MENA, and others have higher populations or larger economies. However, no other region of the world includes countries with such high rates of population growth, and with such high aspirations for a better standard of living, on the basis of so little water. An only slightly imprecise mnemonic is that MENA has five per cent of the world's population but only one per cent of its fresh water. As shown by a host of reports over the past decade, fresh water remains the natural resource of greatest concern to the bulk of the region's people, and by far the most important natural resource for their livelihoods (UNESCO, 2003; CEDARE, 2004; Millennium Ecosystem Assessment, 2005a; HDR, 2006; World Bank, 2007).

Water scarcity is of course not new to MENA. What is new is the rate at which scarcity is becoming worse (HDR, 2006). Per capita renewable fresh water fell from 4000 m^3 per year in 1950 to 1100 m^3 today, and projections indicate that it will drop to 550 by 2050. Anything below 1000 cubic metres per capita per year is commonly taken as indicating physical (first order) water scarcity. However, it is not physical scarcity that determines the level or privation. To the contrary, physical scarcity can be mitigated by institutions that are capable of ensuring that water is extracted in ways that are ecologically sustainable, used in ways that are economically efficient, and distributed in ways that are socially equitable (Wolfe & Brooks, 2003; Varis & Tortajada, 2007; Cook *et al.*, 2009; Molle, 2009). To quote from the Human Development Report (HDR) (2006, p. 3):

> There is more than enough water in the world for domestic purposes, for agriculture and for industry... scarcity is manufactured through political processes and institutions that disadvantage the poor.

The thesis of this paper is that water-related institutions in MENA are slowly changing in ways that seem likely to improve the situation. Notably, water institutions in MENA have gradually been turning their attention from technical and engineering ways to cope with chronic water shortages to political and managerial ways. Three of the most formative tendencies are:

- demand side approaches;
- stakeholder involvement; and
- pro-poor efforts to achieve equity as well as efficiency.

These changes are long overdue and still far from complete, but they are promising for the longer term future of fresh water governance in MENA. Sadly, they are not very different from the Action Plan that emerged from the Mar del Plata Inter-Governmental Conference on Water in 1977 (Biswas, 1988). Had MENA nations brought those recommendations even partially into policy, the region would today be much better placed with respect to both human and ecological needs for water.

Regional Characteristics and Water Management

MENA varies as much in socio-economic and demographic characteristics as in rainfall. Averages have little explanatory value. For example, the Human Development Index value for the Arab States is almost identical to the average for all developing countries.[2] However, because the Arab states include high-income, oil-producing countries, one can infer that the bulk of the countries are below the average. In fact, most rank between 100th and 150th in the list of 177 countries. Worldwide, the region is the second poorest (after South Asia). Selected socio-economic indicators for Lebanon (ranking 78th in the HDR index), Jordan (86th), Egypt (111th), Morocco (123rd) and Yemen (150th) appear in Table 1.

High degrees of urbanization appear to be linked to higher rates of per capita income, greater female literacy, and improved water services. However, the relationship is not direct, nor does it extend consistently to all indicators. A better indicator of the source of poverty may be found in data that focus on the role of agriculture in the national economy (see also Varis & Abu-Zeid, 2009; and the March 2009 issue of *Water International*). Only a few MENA nations derive even one-fifth of their gross domestic product from agriculture (Table 2), but those that do are among the poorest in the region. Further

Table 1. Socio-economic indicators in selected MENA nations

INDICATOR *(TABLE NO.)*	Lebanon	Jordan	Egypt	Morocco	Yemen
Population growth (%/a) (1975–2004) *(5)*	1	3.6	2.1	2.1	3.7
GDP per capita (PPP US$) *(14)*	5800	4700	4200	4300	879
Ratio of female to male earned income *(25)*	0.31	0.30	0.23	0.25	0.30
Urban population (% of total) *(5)*	87	82	43	58	27
Adult literacy rate (male/female) *(24)*	High	95/85	83/60	66/40	72/33
% of population earning less than US$2 per day *(3)*	Low	7	44	14	45
Per capita spending on health (PPP US$) *(5)*	730	440	235	218	89

Note: Italicized numbers in column one show the table where that index appears in the report.
Source: HDR, 2006.

discussion on the combined effects of the region's relatively high rates of population growth and slow rates of income growth and investment can be found in Varis & Abu-Zeid (2009).

Institutional Design

National water management agencies in MENA are, for the most part, oriented to large-scale, engineering-dominated supply systems for urban water supply and for irrigation, a traditional approach that goes back to ancient Sumer and that has, until recently, worked reasonably well. However, as watersheds come to be fully allocated, and as costs of additional supply continue to mount, its deficiencies become evident:

- Public involvement in decision making is rare, and the agencies tend to be insensitive to local practices, gender concerns, and environmental impacts.
- Water is allocated primarily to users who produce revenue, which in rural areas means that irrigation is favoured over household use, and commercial farming (especially for export crops), over (generally smaller) farms producing for local markets.
- Most MENA countries have separate agencies for irrigation water and for drinking water, with no clear guidelines as to how they should collaborate or coordinate

Table 2. Percentage of total water use used for agriculture compared with contribution of agriculture to the economy (selected countries)

Country	Percentage of water used in agriculture	Percentage of GDP obtained from agriculture
Ethiopia	86	55
Sudan	94	33
Egypt	86	16
Yemen	92	16
Morocco	92	14
Turkey	72	14
Jordan	75	6
Lebanon	68	2

Source: Beaumont, 2002.

their efforts. Other uses of water remain statistically small and are either, as with industry, the responsibility of the urban water agency or, as with fishing, fall to ministries with lesser status.

- National and municipal water institutions treat demand management as a secondary responsibility, if they see it as their responsibility at all. Apart from water pricing, they typically implement only those measures that affect their own revenue streams, and ignore water use by the people they serve.

Despite the dominance of centralized management of water in MENA, some forms of decentralization are beginning to appear. Algeria, Jordan and Morocco have built river basin agencies, and Lebanon, Tunisia, and Yemen have devolved water management to sub-national units (World Bank, 2007). As well, in some countries various forms of traditional clan or family control over specific wells or springs persists (for example, Trottier, 1999, on Palestine).

Role of Agriculture

As indicated by Table 2, agriculture accounts for at least two-thirds and typically four-fifths of water use in MENA. No other region of the world uses so large a share of its water for irrigation as does MENA, yet MENA is the most water-stressed region of the world. Of course, the role of agriculture in MENA is as much social as economic. It provides employment for a growing body of unskilled labour and a source of subsistence that might not otherwise exist. Nevertheless, the share of water used for irrigation remains a concern. Typical irrigation practices in MENA use 10,000 cubic metres of water per year for every hectare. That same quantity of water would meet the basic requirements for 1000 people. Clearly, pressure to shift water from agricultural to urban uses is bound to increase.

Agriculture requires special attention for other reasons as well. In urban areas, institutional change affects a critical but small portion of people's lives and incomes. (Exceptions must be made for people engaged in urban agriculture and for local water-dependent industries such as brewing.) Land-use decisions can be separated from water-use decisions, and the balance between water for growing food and water for ecosystem services tends to be complementary. Further, there is a demonstrated willingness in urban areas to pay for better water services, and one can more easily adopt conventional economic approaches to pricing and allocation. None of these characteristics applies to rural areas. In rural areas, water availability and livelihoods are almost the same thing; land-use decisions must be coupled with water-use decisions; and there is a growing conflict between water to grow food and water to protect ecological services (Warner *et al.*, 2006). Full cost pricing of irrigation water is particularly difficult to achieve (Hellegers & Perry, 2006). As a result of these differences, institutional change must be introduced much more cautiously in rural areas.

Urban Water Use and Institutions

Despite their relatively low ranking on the overall Human Development Index, MENA nations do reasonably well in providing drinking water to their people. Even in the poorest nations, two-thirds of the population had "sustainable access to improved water sources" in 2004 (HDR 2006, Table 7), and many nations have rates between 90 and 100% coverage. These data are likely biased toward urban users, and the World Bank (2007) reports that they reflect constructed infrastructure rather than service actually delivered, which is typically lower. Availability of "sustainable access to improved sanitation" is more limited. Only Algeria, Israel, Jordan, Lebanon, Libya, and Syria, plus the oil-rich states, have rates above 90%, and many MENA nations have rates below 50%.

Urban water institutions are typically either quasi-independent agencies or part of an urban affairs ministry. Institutional patterns do not vary much from one country to the next except in the balance between roles for the public and private sectors, which varies from nil in most countries to almost complete in some cities of the Maghreb. In either case, the starting points are (a) providing some basic level of water service for all residents regardless of ability to pay, and (b) treating water as an economic good for the bulk of the population and for commercial and industrial clients. Apart from those who consume just a little water and pay only the 'social tariff,' water rates are moving toward covering full (including capital) costs of supply and wastewater removal. The main problem is less about the inability or unwillingness to pay than poor service delivery and poor quality of water, which creates a vicious circle of non-payment of bills and thus even worse service.

Rural Water Use and Institutions

The starting point for rural water institutions is not so much the value of water as the national plans for agriculture and rural development. In many MENA countries, a key agricultural goal has been self-sufficiency in food. As Varis and Abu-Zeid emphasize (2009), the remarkable thing is not that they have failed to reach this goal, but that given rapid population growth, per capita production of food has not fallen further behind. With the exception of Yemen, most countries are now supplying 3000 calories per person to the bulk of their population. However, evidence suggests that governments continue to rely on rather simplistic notions about the linkages among agriculture, water and rural poverty (World Bank, 2007; Cook, *et al.*, 2009). Certainly, as HDR 2006 indicates (p. 175), the strength of the linkage is "conditioned by a wide range of institutional factors, including efficiency and equity in land distribution".

Future directions for agriculture in MENA are unclear. Some evidence suggests that the region is shifting its main agricultural goal from self-sufficiency to trade, which of course raises other issues (Varis & Abu-Zeid, 2009). It is now clear that farmers do react to increased prices for water by shifting crops and adjusting farming practices (Brooks *et al.*, 2007). However, the gains in economic efficiency can result in higher prices for food, which can in turn stimulate a renewed emphasis on food self-sufficiency (World Bank, 2009). Some of the pressure to achieve food self-sufficiency is relieved by the huge regional imports of grain, and thus the heavy use of virtual water (Allan, 2001), that is, the water used in other nations to produce the grain. Certainly the debate about a policy of food self-sufficiency is very much alive, and the choice to import is anything but politically neutral (Roth & Warner, 2008; World Bank, 2009).

Catalysts for Institutional Change

Water governance in the Middle East and North Africa is changing in response to forces that originate both inside the water sector, as with closing river basins, and outside it, as with population growth (Saleth & Dinar, 2005). Given some role for each set of forces, it is futile to try to resolve the region's water problems by working entirely within the water sector. Varis & Abu-Zeid (2009) argue that economic factors weigh more heavily than non-economic one in bringing about change, and we tend to agree. However, we agree fully with analysts who argue that the critical forces have important political components, and that they cannot be resolved by technical responses alone (World Bank, 2007; Cook, 2009; Molle, 2009).

This section describes the main internal and external forces that appear to be driving institutional change in fresh water management in MENA. The role of conflict, which does not fit neatly into this framework, is discussed in a third sub-section.

Internal Forces

Of all the forces leading to revised thinking about institutional design and operations in MENA, probably most important is recognition that most MENA nations are approaching—or, indeed, have already passed—the end of viable additions to their renewable water supplies. Whole river basins are 'closing' in the sense that there are no further dam sites with benefit–cost ratios even close to 1, and no aquifers that can be tapped in accessible locations. Such recognition has by no means reversed the traditional attention to supply management, nor has it permeated deeply into water sector agencies, but it is stimulating senior management to think about ways to live within natural limits.

A parallel influence within MENA is a general thrust toward democratization—not just decentralization—of natural resources management. To now, participatory management for water has been restricted to the lowest level of irrigation canals. Initiatives are now being taken to extend participation upward to manage branch and feeder canals, and also to build watershed management institutions that involve a range of stakeholders in policy development and even, in some cases, operations and management. Though they typically fall short of involving all stakeholders, these initiatives are worth encouraging for their potential to improve both water-use efficiency and water-use equity.

Except in the oil producing countries, governments in MENA are increasingly budget constrained. One result has been the search for areas where government responsibilities can be devolved to private agencies, public utilities and cooperatives. The need to cut government expenditures has been the single greatest reason for wider acceptance of participatory irrigation management under which water user groups are expected to find ways to cover their own operation and maintenance costs from the beneficiaries of the irrigation water (Brooks *et al.*, 2007). The same issue has led to more attention to increases in the price of water, particularly in urban areas, which in turn promotes greater effort at water conservation.

Though the drive toward more financial accountability is less evident for agriculture, it does appear in three ways. First, as indicated just above, water users associations are being induced to cover operations and maintenance costs for irrigation canals. Second, some countries designate regions for private participation in the financing of irrigation infrastructure. Such approaches are only applicable to commercial farms, but they have the concomitant effect of promoting the introduction of modern irrigation equipment so that water-use efficiency is higher than in other areas. Third, MENA nations are increasingly reluctant to go into debt to obtain the huge amount of capital required for new irrigation systems. This is a convenient shift given the parallel reluctance of international banks and donor agencies to lend money for such projects.

Concerns for public health, particularly in rapidly expanding urban areas, are coming to play a larger role for governments across MENA. As would be expected, many of the solutions require attention to water. For example, in 2002 diarrhoea caused three times as many deaths per million people in the less affluent parts of MENA than in Latin America, despite comparable levels of per capita income (World Bank, 2007). However, it is not so much high disease rates that stimulate action as sudden peaks in those rates. Cities that have faced outbreaks from inadequate treatment of water, as with Aleppo and Amman, typically not only upgrade their water systems but also reform the institutions mandated to deliver potable water. Similarly, the desire to avoid health impacts and the need to deal with growing volumes of wastewater have led urban water utilities in many MENA nations to develop institutions to promote and regulate re-use of treated wastewater (Scott *et al.*, 2004; Brooks *et al.*, 2007).

External Forces

The most evident external force that will directly affect the availability of water in MENA is global climate change. Different scenarios imply different results for many parts of the world, but almost all models project higher temperatures, lower rainfall, and longer droughts for MENA. Even if such models remain uncertain in detail, it would be foolish to ignore them. Moderate to severe socio-economic effects can be anticipated in countries across the region, regardless of per capita income (World Bank, 2007), but those effects can also be expected to have disproportionately adverse effect on poorer nations and poorer people (Pachauri, 2004). Just as worrisome, some analysts suggest that there are good reasons to expect increased and more violent conflict in MENA (Brown & Crawford, 2009). Some sectors, and notably agriculture, on which the bulk of livelihoods in MENA depend, are also threatened, particularly in rain-fed fields but also for areas under irrigation (Devereux & Edwards, 2004; Mendelson, 2009).

MENA is not responsible for the greenhouse gas emissions that are responsible for the bulk of global climate change, but it seems destined to suffer the worst effects. Defensive strategies are clearly needed, as are more coordinated actions among MENA nations, including those that do not have particularly good relations with one another. Professional opinions vary on exactly what to do and when it should be done, but a notably useful overview, presented in a journalistic style, appears in *Bitter Lemons International* (Edition 34 Volume 8 – 03 September 2009.[3]

Working in the opposite direction are technological developments that can improve efficiency, and to some degree equity and sustainability, in the use of water. Among other things, gains can be obtained through better ways to detect and repair leaks in the typically very leaky water systems of cities in MENA. Drip (micro) irrigation is also coming to be more common. Some developments, such as sensors that control valves so plants get water only when and as much as they need, are only appropriate for large commercial farms; other developments are identifying ways for drip irrigation to be used by small farmers without much capital (Postel *et al.*, 2001). The key force is not new research itself, but the institutions that direct research and that effectively determine whom the research will benefit.

Finally, more attention is being placed today on human rights and social and gender equity. About a third of the general tables presented at the end of the 2006 Human Development Report focus directly on social or gender equity, with nations implicitly criticized on the extent to which they fail on one or another of the indicators of equity. The same thrust is also reflected in the *General Comment on the Right to Water* from the United Nations, Sub-Committee on Economic, Social and Cultural Rights (a part of the UN Commission on Human Rights) (2002). The Committee asserts that rights to water exist on the basis that it "... is indispensable for leading a life in human dignity. It is a prerequisite for the realization of other rights."

Though limited mainly to household water, the growing emphases on human rights and on gender are gradually exerting an influence on water management goals, policies and programmes across MENA (Salath *et al.*, 2003; Klawitter & Qazzaz, 2005; Biswas *et al.*, 2008). Rather than assuming that more and cleaner water improves everyone's life equally, governments will be forced to show how their efforts are helping those who, till now, have been disadvantaged in access to water. The water-poverty-gender nexus in MENA is a fruitful field for further research.

Conflict and Institutional Reform

Though few careful analysts really believe that the next war in MENA will be over water, there will be no shortage of conflict. The relevance to the subject of this article stems from the tendency of institutional reform to be conflict adverse. Reform tends to slow down, if not disappear entirely, in areas under conflict. Per contra, the period when tensions start to relax appears to be a propitious time for institutional change. It is a time when it becomes feasible to negotiate the creation of basin-wide international management authorities, and to review traditional patterns for allocating water among competing users and uses. The best example of such a change occurred after the collapse of the old South African government, which favoured a top-down, supply-oriented approach to water and allocated most water to white farmers. The African National Congress established constitutional guarantees of water (for people and for the environment), and it changed governance strategies in many ways including, for example, giving management authority to those who work the land as opposed to those who own it (Schreiner *et al.*, 2002).

An attenuated example of the same potential was evident during the post-Oslo period of relaxed tensions between Israel and Palestine. Not only were gains achieved in both the Water Resources and the Environment Working Groups of the Multilateral Track of the Peace Process, but a range of "second track" (non-diplomatic) programmes were initiated. Led by academic institutions and non-governmental organizations (NGOs), they delved into areas that governments had avoided until that time. Notable were investigations for bilateral Palestinian-Israeli management of the Mountain Aquifer (Feitelson & Haddad, 2000) and for trilateral Palestinian-Jordanian-Israeli management of the Dead Sea basin (Bromberg, 2004), both of which were supported by the International Development Research Centre (IDRC) and other donor agencies. New governmental institutions were also created. However, as Israeli occupation of the West Bank and, until recently, Gaza continued, traditional forms of water management re-exerted themselves, as did Israeli use of various forms of power other than direct military intervention (Trottier, 2007; Zeitoun, 2008). The dynamism of the Oslo period has abated, and it is not clear what water institutions will serve Palestinians in the future.

Resultant of Forces

All MENA nations continue to work within highly centralized water management institutions, and, given the importance of water to their economic and social development, such centralization is probably necessary. One happy result of this centralization, and of the forces described in this section, is a new focus on those strategic issues for which clear directions and management are most necessary. The character of the shift from top-down management to strategic policy development is unique to each country but typically involves the creation of a water planning document that looks 20 or 30 years into the future and that is cast within the framework of integrated water resources management (IWRM). Though IWRM has been criticized because of the enormous difficulties of implementation (Biswas, 2004; Saravanan *et al.*, 2009), it is nevertheless a useful way to stimulate deeper thinking about water policy. Among the gains found in many MENA nations are the following:

- publication (or ongoing preparation) of national water strategies;
- presentation of IWRM as the core element in that strategy;
- treatment of water demand management and of water quality as full complements to water supply in IWRM; and
- insistence on greater inter-ministerial cooperation in furtherance of the strategy.

The process is most evident in Egypt, Jordan, Morocco and Tunisia (see, for example, Government of Jordan, 2009. Efforts are less advanced in other Arab countries, but almost everywhere there is attention to developing strategies that will accommodate water demand to hydraulic limitations. The experience of Morocco, which has undergone the most thorough reform of its water institutions, is encouraging. According to Doukali (2005, p. 87):

> the reform experience of Morocco suggests that, although undertaking initial reform can be difficult, subsequent reforms are relatively easier as the country consolidates and adjusts with the earlier reforms.

Of course, barriers to implementation of institutional reform remain strong. Though water agencies may now be mandated to work within a framework of IWRM, they are still staffed by the same people who have spent their careers focusing on supply management. It is hardly likely that they will promote demand-side approaches, local management, and environmental protection with the enthusiasm (and the budget) they require.

Gains and Gaps in Institutional Change

This section presents highly selected examples to show where institutional change is gaining ground, and where progress is lacking.

Where Gains Are Being Made

Perhaps the most common adjustment to the failures of top-down management to achieve its objectives was the creation of Water Users Associations (WUAs) to enable farmers to participate in the operations and maintenance of local irrigation systems (among many others: Attia, 2005; Yildirim & Cakmak, 2004; Salman *et al.*, 2008; and Supplement 3 to Volume 58 of *Irrigation and Drainage*, 2009). Water-use efficiency typically increases by 30 to 50 percent with WUAs, and the energy used for pumping is cut in half or even more

(Attia, 2005; Doukali, 2005). The increase in water efficiency does not necessarily yield a reduction in water use; more commonly, it means that tail-enders on the water system now get water regularly, that is, greater equity and efficiency, but not less water use. Women say they too get a better break inside WUAs, and there are other less well-documented benefits, such as reduction in conflict and improvements in family health (Van Hoffwegen, 2003). On the other hand, WUAs are no panacea. Richer farmers tend to benefit most, and efforts to extend them from tertiary to secondary (feeder) canals have not been so successful as hoped (Brooks *et al.*, 2007; Cook, 2009).

Many MENA nations revised their water valuation systems within the last decade, and most of them did so on the basis of studies showing the advantages of marginal cost pricing. Some incorporated smarter pricing as well. For example, Jordan and Tunisia prevent richer households from benefiting from the social tariff by requiring consumers who use more water than that covered by the initial block to pay at the higher rate for the full, not the marginal, volume.

A long-standing bias against investing in agricultural research in less favoured areas appears to be changing. Despite the dominance of irrigation as a use for water, three-fourths of all farmland in Arab nations depends exclusively on rain (HDR 2006, p. 177). (Egypt, Israel, Libya and the nations of the Arabian Peninsula are the main exceptions.) Productivity on rain-fed farms is far below that for irrigated farms, but they offer great potential for improvement, which in turn has potential for alleviating poverty (IWMI, 2007). As one example, three-fourths of all wheat in MENA is grown under rain-fed conditions with typical productivity of one tonne per hectare, a rate that can certainly be doubled and possibly quadrupled by a variety of techniques to store water and protect soil moisture through conservation tillage and, in some areas, supplementary irrigation (IWMI, 2007, Policy Action 5).

Given the statistically significant relationship between greater rainfall variability and lower per capita GDP (Brown & Lall, 2006), the roughly proportionate reduction in rural poverty with every increase in agricultural-water productivity (Thirtle *et al.*, 2002), and indications from Asia of a positive relationship between agricultural productivity and equity (Hussain, 2005), it is evident that there is much to be gained from research on pro-poor strategies for rural water management (see IWMI, 2007, Policy Action 2; HDR, 2006). With a higher proportion of women working in rain-fed than in irrigated agriculture, women's poverty can be alleviated with the same increase (Van Koppen, 1999; Gender & Water Alliance, 2003; Singh *et al.*, 2005; IWMI, 2007). However, Pender & Hazell (2000, p. 2) caution:

> No single strategy will work in all less-favored areas. . . . The key is to identify and implement the appropriate portfolio of such public and private investments for different circumstances in less-favored areas. Achieving this goal requires more than simply new technologies or policies. It requires responsive and effective institutions to mobilize such investments and to ensure accountability, efficient management, and equitable distribution of benefits.

Where Gains Are Not Being Made

Important gaps remain in shifting water management institutions in MENA toward goals of equity, efficiency and sustainability. Three of the particularly important gaps in areas fully within the capacity of the water sector to correct (as, notably, pro-poor strategies are not) follow:

Demand management institutions. Though demand management is slowly gaining attention in water management, few institutions have the mandate to develop and implement programmes. The naive assumption in most countries is that the same agencies that have been delivering supply management all these years will now devote equal attention to demand management. Worse yet, in many cases, drinking water utilities in MENA countries explicitly disavow any interest in or responsibility for the efficiency of water use beyond the main valve—inside the house or building or factory. Experience with the parallel shift from supply to demand management for energy shows that committed agencies with explicit responsibility for developing demand management policies and with direct support from senior management are essential (Wolfe & Brooks, 2003; Brooks & Wolfe, 2007).

Ecological demand for water. Most states in MENA treat their water resources as if all water can be extracted for human uses. There is little recognition that a large share of the water must be left in place to provide services ranging from sanitation to flood control and habitat protection (Postel & Thompson, 2005; Falkenmark & Rockstrom, 2004; IWMI, 2007). Even approximate numbers suggest that intact ecosystems provide economic values for society well above the private values achieved after land is converted to purportedly "more productive" uses (Millennium Ecosystem Assessment, 2005b). Yet few MENA nations have started to use the tools or adopt the management strategies that are available (IWMI, 2007, Chapter 6) for ensuring that there is enough water for both agriculture and the environment.

Groundwater management. Groundwater management continues to be neglected in most of MENA. According to the World Bank (2007), over-extraction of groundwater is undermining national assets at rates equivalent to 1–2% of GDP every year in some countries of the region. Even in the better managed systems, political considerations play an excessive role in groundwater management for many years before reforms are implemented (Feitelson, 2005). Ground water must be seen as a public, not a private, asset, and a government body must have clear responsibility for allocating permits to drill and to withdraw water, and to promote aquifer recharge, even on private land.

Conclusion: From Board Room to Homes and Fields

In response to the catalytic forces identified in this paper, institutional change to improve water governance is occurring in the Middle East and North Africa (MENA). However, it is not occurring to the extent nor with the speed that is required by the evident contrast between growing demands for water and distinctly limited supplies. Many of the reasons for the limited extent and slow rate of institutional change are found inside the water sector itself. The problem is less the lack of good policy design than the combination of ineffective policy implementation and lack of incentives for institutional change (Brooks & Wolfe, 2007; Varis & Tortajada, 2007). Almost everywhere the needed institutional changes find support among senior planners and in the minister's cabinet, but those changes are not carried through to lower levels of the organization. At least two forces are needed to institute a climate for institutional change:

The first needed force is identification of political and administrative 'champions' who have the courage and the status to mobilize and support a process of change within existing institutions (Molle, 2009). Bureaucracies do not move easily, particularly not when the required change will reverse long-standing policies. However, they will move when they are given clear instructions and when they are sure that those instructions are backed by senior managers, not just by planners in a section far removed from daily operations.

The second needed force is a long-term effort toward capacity development for institutional change (Wolfe & Brooks, 2003; Brooks & Wolfe, 2007). A few have begun to emerge in the last few years. Programmes established by IDRC[4] and by InWEent (Varis & Tortajada, 2007) are very promising, but results are only now beginning to appear. In the absence of leadership from graduates of these programmes, it is unlikely that MENA nations will have that mix of knowledge and of political that will be necessary for sustained improvement in water governance.

A third force, non-governmental organizations, may become an important force in the future, but they are still nascent in most MENA nations (World Bank, 2007). Only in Israel and Palestine, and to a lesser extent, Jordan, Lebanon and the three countries of the Maghreb, do NGOs play a role in water policy.

Where does MENA stand at this time with respect to improved governance of fresh water? Generalizations are difficult, given the region's size and diversity. However, the following are suggested:

- There is almost universal recognition of the need for institutional change in the water sector, including a commitment to integrated water resources management.
- For the most part, the kinds of institutional change that are needed are known, though application in the specific circumstances of each country is varied.
- There is, as yet, only limited evidence of the political will to ensure that the institutional change is as thorough as it needs to be.
- Till now, greater changes have been effected in urban areas than in rural.
- Of the three goals for new institutions mentioned at the start of this paper—greater attention to demand management, stakeholder participation, and pro-poor strategies—modest gains are evident on the first two, but comparatively little on the third. The chapter on poverty reduction in IWMI's Comprehensive Assessment of Water Management in Agriculture has nearly 150 reference notes; not one refers specifically to MENA (IWMI, 2007).

The World Bank's Flagship Development Report states, "water need not be a constraint to economic development and social stability in MENA" (World Bank, 2007, p. xxvii). However, that conclusion is heavily dependent not just on planning reforms of water sector institutions but also on their effective implementation. Again citing the World Bank report, reform "will involve some difficult choices and painful changes" (*Ibid.*, p. xxviii). The best argument for pursuing the reforms is that, in their absence, the choices will become even more difficult and the changes even more painful year by year.

Notes

1. This paper was originally prepared for a workshop in Zaragoza, Spain, organized by the Third World Centre for Water Management and the International Centre for Water, to celebrate and build upon the intergovernmental water meeting that was held in Mar del Plata in 1987.

2. The Human Development Index is intended to be "a summary measure of human development" (HDR 2006, p. 394). It is calculated as the average of three indices: 1) life expectancy at birth; 2) adult literacy combined with total school enrollment at two-thirds/one-third weighting; and 3) gross domestic product per capita.
3. Available at: http://www.bitterlemons-international.org/previous.php?opt=1&id=288.
4. http://www.idrc.ca/wadimena/ev-96990-201-1-DO_TOPIC.html, and http://www.idrc.ca/wadimena/ev-138409-201-1-DO_TOPIC.html

References

Allan, J. A. (2001) *The Middle East Water Question: Hydropolitics and the Global Economy* (London: I. B. Tauris).

Attia, B. (2005) Decentralization and participatory irrigation, in: E. Baroudy, A. A. Lahlou & B. Attia (Eds) *Managing Water Demand: Policies, Practices, and Lessons from the Middle East and North Africa Forums* (London: IWA Publishing and Ottawa: IDRC Books).

Beaumont, P. (2002) Water policies for the Middle East in the 21st century: the new economic realities, *International Journal of Water Resources Development*, 18(2), pp. 315–334.

Biswas, A. K. (1988) United Nations water conference: implementation over the past decade, *International Journal of Water Resources Development*, 4(3), pp. 148–159.

Biswas, A. K. (2004) Integrated water resources management: a re-assessment, *Water International*, 29(2), pp. 248–256.

Biswas, A. K., Rached, E. & Tortajada, C. (Eds) (2008) *Water as a Human Right for the Middle East and North Africa* (London: Routledge, Ottawa: IDRC).

Bromberg, G. (2004) Water and peace: for clues to resolving the Middle East conflict, consider the case of the embattled Dead Sea, *World-Watch*, 17(4), pp. 24–30.

Brooks, D. B., Thompson, L. & El Fattal, L. (2007) Water demand management in the Middle East and North Africa: observations from the IDRC forums and lessons for the future. *Water International*, 32(2), pp. 193–204.

Brooks, D. B. & Wolfe, S. (2007) *Institutional Assessment for Effective WDM Implementation & Capacity Development*, Water Demand Management Research Series (Cairo, Egypt: IDRC). Available at: http://www.idrc.ca/wadimena

Brown, C. & Lall, U. (2006) Water and economic development: the role of variability and a framework for resilience, *Natural Resources Forum*, 30(4), pp. 306–317.

Brown, O. & Crawford, A. (2009) *Rising Temperatures, Rising Tensions: Climate Change and the Risk of Violent Conflict in the Middle East* (Winnipeg, Canada: International Institute of Sustainable Development).

CEDARE (Centre for Environment and Development for Arab Region and Europe) (2004) *State of the Water in the Arab Region* (Cairo: Arab Water Council & Centre for Environment and Development for Arab Region and Europe).

Cook, S. E., Fisher, M. J., Andersson, M. S., Rubiano, J. & Giordano, M. (2009) Water, food and livelihoods in river basins, *Water International*, 34(1), pp. 13–29.

Devereux, S. & Edwards, J. (2004) Climate change and food security, *IDS Bulletin*, 35(3), pp. 22–30.

Doukkali, M. R. (2005) Water institutional reforms in Morocco, *Water Policy*, 7(1), pp. 71–88.

Doukkali, M. R. (2005) Water institutional reforms: theory and practice, *Water Policy*, 7(1), pp. 1–19.

Falkenmark, M. & Rockström, J. (2004) *Balancing Water for Humans and Nature: The New Approach to Ecohydrology* (London: Earthscan Publications).

Feitelson, E. (2005) Political economy of groundwater exploitation: the Israeli case, *International Journal of Water Resources Development*, 21(3), pp. 413–423.

Feitelson, E. & Haddad, M. (2000) *Management of Shared Groundwater Resources: The Israeli-Palestinian Case with an International Perspective* (Ottawa: IDRC Books, Dordrecht, Netherlands: Kluwer Academic Publishers).

Gender and Water Alliance (2003) *Gender Perspectives on Policies in the Water Sector*. The Gender and Water Development Report 2003, Leicestershire, UK: Loughborough University, WEDC.

Gleick, P. (2000) The human right to water, in: P. Gleick (Ed.) *The World's Water: 2000–2001* (Washington, DC: Island Press).

Government of Jordan (2009) *Water for Life: Jordan's Water Strategy 2008–2022* (Amman: Ministry of Water and Irrigation).

Hellegers, P. J. G. J. & Perry, C. J. (2006) Can irrigation water use be guided by market prices? theory and practice, *International Journal of Water Resources Research*, 22(1), pp. 79–86.

Human Development Report (HDR) (2006) *Beyond Scarcity: Power, Poverty and the Global Water Crisis* (New York: United Nations Development Programme).

Hussain, I. (2005) *Pro-Poor Intervention Strategies in Irrigated Agriculture in Asia: Final Synthesis Report* (Colombo, Sri Lanka: IWMI).

International Water Management Institute (IWMI) (2007) *Water for Food, Water for Life: Comprehensive Assessment of Water Management in Agriculture* (London: Earthscan, Colombo, Sri Lanka: IWMI).

Klawitter, S. & Qazzaz, H. (2005) Water as a human right: the understanding of water in the Arab countries of the Middle East, *International Journal of Water Resources Development*, 21(2), pp. 253–271.

Mendelson, R. (2009) The impact of climate change on agriculture in developing countries, *Journal of Natural Resources Policy Research*, 1(1), pp. 5–19.

Millennium Ecosystem Assessment (2005a) *Ecosystems and Human Well-Being: Current State and Trends*, Vol. 1, Chap. 7: Freshwater ecosystem services (Washington, DC: World Resources Institute).

Millennium Ecosystem Assessment (2005b) *Ecosystems and Human Well-Being: Wetlands and Water-Synthesis* (Washington, DC: World Resources Institute).

Molle, F. (2009) Water, politics and river basin governance: repoliticizing approaches to river basin management, *Water International*, 34(1), pp. 62–70.

Pachauri, R. K. (2004) Climate change and its implications for development: the role of IPCC assessments, *IDS Bulletin*, 35(3), pp. 11–14.

Pender, J. & Hazell, P. (2000) *Promoting Sustainable Development in Less-Favoured Areas: Overview* International Food Policy Research Centre Policy Briefs on Promoting Sustainable Development in Less-Favored Areas, Focus 4, Brief 1, Washington, DC: IFPRI.

Postel, S., Polak, P., Gonzales, F. & Keller, J. (2001) Drip irrigation for small farmers: a new initiative to alleviate hunger and poverty, *Water International*, 26(1), pp. 3–13.

Postel, S. & Thompson, B. H. (2005) Watershed protection: capturing the benefits of nature's water supply services, *Natural Resources Forum*, 29(2), pp. 98–108.

Rosegrant, M. W., Cai, X. M. & Cline, S. A. (2002) *World Water and Food to 2025: Dealing with Scarcity* (Washington, DC: IFPRI).

Roth, D. & Warner, J. (2008) Virtual water: virtuous impact? the unsteady state of virtual water, *Agriculture and Human Values*, 25(2), pp. 257–270.

Saghir, J. (2002) Reflections on water pricing and tariff design: key principles. Paper presented at the Water Valuation Forum, Beirut, 26 June 2002. Available at: http://www.idrc.ca/en/ev-44056-201-1-DO_TOPIC.html

Saleth, R. M. & Dinar, A. (2005) Water institutional reforms: theory and practice, *Water Policy*, 7(1), pp. 1–19.

Salman, A., Al-Karablieh, E. & Haddadin, M. (2008) participatory irrigation water management in the Jordan Valley, *Water Policy*, 10(4), pp. 305–322.

Saleth, R. M. & Dinar, A. (2000) Institutional changes in global water sectors: trends, patterns and implications, *Water Policy*, 2(3), pp. 175–199.

Saleth, M., Samad, M., Molden, D. & Hussain, I. (2003) Water, poverty and gender: an overview of issues and policies, *Water Policy*, 5(5/6), pp. 385–388.

Saravanan, V. S., McDonald, G. T. & Mollinga, P. P. (2009) Critical review of integrated water resources management: moving beyond polarised discourse, *Natural Resources Forum*, 33(1), pp. 76–86.

Schreiner, B., van Koppen, B. & Khumbane, T. (2002) From bucket to basin: a new paradigm for water management, poverty eradication and gender equity, in: A. Turton & R. Henwood (Eds) *Hydropolitics in the Developing World: A Southern African Perspective* (Pretoria, South Africa: African Water Issues Research Unit, University of Pretoria).

Scott, C. A., Faruqui, N. I. & Raschid-Sally, L. (2004) *Wastewater Use in Irrigated Agriculture: Confronting the Livelihood and Environmental Realities* (Oxford, UK: CAB Publishing).

Singh, N., Jacks, G. & Bhattacharya, P. (2005) Women and community water supply programmes: an analysis from a socio-cultural perspective, *Natural Resources Forum*, 29(3), pp. 213–223.

Thirtle, C., Beyers, L., Lin, L., McKenzie-Hill, V., Irz, X., Wiggins, S. & Piesse, J. (2002) *The Impacts of Changes in Agricultural Productivity on the Incidence of Poverty in Developing Countries*, Department for International Development Report No. 7946. London, UK: DFID.

Trottier, J. (1999) *Hydropolitics in the West Bank and Gaza Strip* (Jerusalem: Palestinian Academic Society for the Study of International Affairs).

Trottier, J. (2007) A wall, water and power: the Israeli "separation fence", *Review of International Studies*, 33, pp. 105–127.

UNESCO, World Water Assessment Programme (2003) *Water for People; Water for Life* (Oxford, UK: Berghahn Books).

United Nations, Sub-Committee on Economic, Social and Cultural Rights (2002) *General Comment on the Right to Water* (New York: United Nations).

Van Hoffwegen, P. (2003) Decentralization and participation: farmers' perspectives and incentives. Paper presented at the IDRC Forum on Decentralization and Participatory Irrigation Management, February 2003, Cairo, Egypt.

Van Koppen, B. (1999) Targeting irrigation support to poor women and men, *International Journal of Water Resources Development*, 15(1/2), pp. 121–140.

Varis, O. & Abu-Zeid, K. (2009) Socio-economic and environmental aspects of water management in the 21st century: trends, challenges and prospects for the MENA region, *International Journal of Water Resources Development*, 25(3), pp. 507–522.

Varis, O. & Tortajada, C. (2007) Water governance in the MENA region: critical issues and the way forward, Feldafing, Germany: Inwent Capacity Building International. Paper presented at an International Conference in Cairo, 23–27 June 2007.

Warner, J. F., Bindraban, P. S. & Keulen, H. (2006) Introduction: water for food and ecosystems—how to cut which pie?, *International Journal of Water Resources Development*, 22(1), pp. 3–15.

Wolfe, S, & Brooks, D. B. (2003) Water scarcity: an alternative view and its implications for policy and capacity building, *Natural Resources Forum*, 27(2), pp. 99–107.

World Bank (2007) *Making the Most of Scarcity: Accountability for Better Water Management Results in the Middle East and North Africa* (Washington, DC: The World Bank, Office of the Chief Economist).

World Bank (2009) *Improving Food Security in Arab Countries*. World Bank Report (Washington, DC: The World Bank). Available from: http://www.reliefweb.int/rw/rwb.nsf/db900SID/KHII-7RC4GY?OpenDocument.

Yildirim, Y. E. & Cakmak, B. (2004) Participatory irrigation management in Turkey, *International Journal of Water Resources Development*, 20(2), pp. 219–228.

Zeitoun, M. (2008) *Power and Water in the Middle East: The Hidden Politics of the Palestinian-Israeli Water Conflict* (London, UK: I. B. Tauris).

[illegible]

[illegible]

[illegible]

[illegible]

[illegible]

[illegible]

[illegible]

[illegible]

[illegible]

[illegible]

[illegible]

[illegible]

[illegible]

[illegible]

Water Supply of Phnom Penh: An Example of Good Governance

ASIT K. BISWAS* & CECILIA TORTAJADA**
*Third World Centre for Water Management, Mexico, and Institute of Water Policy, Lee Kuan Yew School of Public Policy, Singapore; **Institute of Water Policy, Lee Kuan Yew School of Public Policy, Singapore, and International Centre for Water and Environment (CIAMA), Zaragoza, Spain

ABSTRACT *During the 1980s, after years of centralized management and a culture of inefficiency and corruption, the Phnom Penh Water Supply Authority (PPWSA) was dysfunctional. Unaccounted for water routinely surpassed 70%. During the 15-year timeframe between 1993 and 2008, it experienced a metamorphosis; with enlightened management and dedicated and competent staff, the Authority has been completely turned around. Its annual water production increased by 437%, the distribution network by 557%, pressure of the system by 1260%, and its customer base by 662%. Unaccounted for water was reduced from 72% to 6.19%. Its profit increased consistently as has the amount of taxes paid to the Cambodian Government. This paper is a summary of an independent evaluation that examines how this remarkable transformation has been achieved.*

Introduction

Cambodia is well endowed with available water resources. It has a high annual rainfall (up to about 3000 mm in the highlands), three major rivers (Mekong, Bassac and Tonle Sap) and their many tributaries, and excellent sources of groundwater both in terms of quantity and quality. Until the late 1960s, urban water services in Cambodia were similar to what then existed in many of its neighbouring countries. Many of the residents of Phnom Penh had an uninterrupted 24-hour water supply of reasonable quality water.

The situation, however, changed dramatically after the late 1960s due to considerable political turmoil, and this continued unabated for the next two decades. The situation worsened during the four-year rule of the Khmer Rouge who captured Phnom Penh in 1975. The Khmer Rouge attempted to make Cambodia a classless society by forcing people to work in agricultural communes. Its strong ideology included isolating the country from all types of external influences.

The social, political, economic and institutional turmoil took their toll on all of Cambodia's development sectors during the decades of the 1970s and the 1980s. Urban water management was not an exception. All forms of social services in the country, from education to health, became totally dysfunctional, and all urban infrastructure was grossly neglected during these two lost decades. There were no new investments in water or any type of infrastructure, nor was infrastructure maintained and operated at a minimum acceptable level. Skilled personnel were simply not available to properly maintain and operate the existing urban water system, and available records of the urban utilities were destroyed.

By the 1980s, the Phnom Penh Water Supply Authority (PPWSA, the Authority) was in bad shape in institutional, financial and management terms. Years of centralized, top-down management had contributed to a culture of inefficiency, bureaucracy, corruption and conformity. The staff did the least possible work that was needed; it was estimated that 80% of the staff worked less than two hours per day. Salaries were very low, the working environment was poor, and corruption was endemic.

The lack of clean water and poor wastewater management practices took a heavy toll on the people of Cambodia during this period. In the mid-1990s diarrhoeal diseases were endemic in the country, infant mortality rates were some of the highest in the world (115 per 1000 live births), as were the morbidity rates. Water-related diseases accounted for nearly 30% of all hospitalizations. In Phnom Penh, unaccounted for water (UFW) routinely exceeded 80% during the late 1980s, and the pressure of water was so low that most households received no supply of water, even though they were connected to the system.

By the early 1990s, PPWSA as an institution and its overall management were dysfunctional, and as a result consumers received very poor service and a low level of service delivery. In 1992 it had only five engineers, and most of its staff members were incapable even of reading meters accurately. Fortunately, the institution started to change radically from 1993.

Progress since 1994

The United States lifted the trade embargo on Cambodia in 1992, and the Royal Government of Cambodia was established the following year after the successful election supervised by the United Nations. Shortly thereafter, the Cambodia Government requested help from the various multinational and bilateral aid institutions to rehabilitate its battered urban water systems and to formulate a future long term plan.

Immediately after the trade embargo on Cambodia ended, the French provided an aid grant to improve the water distribution network. It was promptly followed with another French grant in 1993 to improve the city's water supply facilities. Also in 1993, the UNDP provided a grant for technical assistance for the rehabilitation of the water utilities of Phnom Penh and Sihanoukville, administered by the World Bank.

Among these early Cambodian requests was one to the Japanese government in early 1993 for technical assistance for the rehabilitation of the existing water supply system of Phnom Penh, as well as to draw up a future water development plan for the city. The Japanese government carried out a study to develop a master plan and feasibility study of the Phnom Penh water supply system. It was conducted by Tokyo Engineering Consultants in association with Nihon Suido Consultants (1993a, 1993b) for the Japan International Cooperation Agency (JICA). The study team arrived in Phnom Penh in February 1993, and within a short period of 10 months completed the final report and delivered it to the PPWSA.

In retrospect, this study turned out to be very important because it served as a road map for the next several years for the PPWSA. The plan was developed in close consultation and discussion with the Authority, and thus it was acceptable to them. This plan became the blue print for the development of the utility over subsequent years. All of the projects supported by the other donors had to fit within this framework and also comply with its overall philosophy. It provided an excellent basis for donor coordination, albeit indirectly. In 1995, JICA provided a grant of $25 million to improve the water supply facilities under Phase I and II, and another $21.326 million grant in 1997 for Phase II. Other major donors who provided financial and technical support to the PPWSA from 1997 included the Asian Development Bank (ADB, 1996) and the World Bank. However, as was the case for the initial financial support, these were loans to the Authority and not grants.

The formulation of this Plan and its subsequent implementation by the PPWSA and foreign donors, as well as the continuing association of JICA on a long-term basis to provide assistance to the PPWSA on different aspects of water supply, including infrastructure development and management and capacity building, have been an important element for the subsequent success of the Authority. However, it was only one of the several other important components which have contributed to the remarkable and steady improvements in the performance of the utility.

Urbanization

In terms of urbanization, Phnom Penh is truly a unique case. Until the late 1960s it was a 'normal' town, coming to grips with decades of French colonial rule and then independence. The population of the city was growing steadily, both due to natural causes and to rural-to-urban migration. In contrast with the other ex-colonial capital cities of Asia and Africa, the growth rate was not spectacular. It was manageable. However, with the Khmer Rouge in power with its strict policy of forcing urban residents to relocate to rural areas for agricultural activities, it became the only city in recent memory, if not in the entire history of mankind, to be massively and immediately depopulated for ideological reasons. Due to Khmer Rouge policies, the population of Phnom Penh in 1979 was estimated to have declined to about 122,800.

With the fall of the Khmer Rouge, the population started to trickle back to Phnom Penh. After a period of time, the rate of return of the former residents increased. When the situation stabilized, the urbanization process, as in other Asia urban centres, gradually accelerated. The population growth of Phnom Penh between 1993 and 2008 is shown in Figure 1. There was an important difference in the population increase of Phnom Penh after the Khmer Rouge era. Many of the original residents of Phnom Penh had been killed. Also, after the traumatic experience, many families decided not to return to Phnom Penh and chose to settle elsewhere in the country. This loss of population was more than made up by the soldiers who were posted in Phnom Penh and who brought their extended families to the city from the interior of the country. These migrants had never lived in an urban setting and were more self-reliant. They managed to live without a regular supply of water as was the case when they were living in rural areas. They neither expected, nor received, water from the PPWSA during the transition period.

In terms of supplying water to the residents and the industrial users of Phnom Penh, the PPWSA faced two critical challenges. First was to restore a reasonable service delivery to a limited geographical area, which was initially considerably smaller than the situation before the Khmer Rouge assumed power. Second was the supply of drinking water to the entrie Phnom Penh city. As the PPWSA started to improve, especially from 1994, its

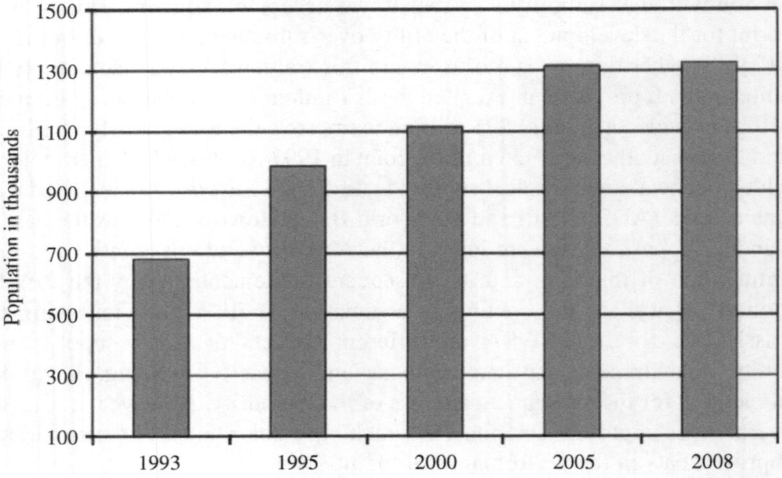

Figure 1. Population of Phnom Penh, 1993–2008

service delivery area constantly expanded and drinkable water delivery increased. Figure 2 indicates the area to which water was provided between 1993 and 2000 primarily because of pipe replacement and the rehabilitation of the overall supply system.

During the 2001–2008 period, and especially after the spectacular success of the PPWSA in providing a clean drinkable water supply on a 24-hour basis, its service area constantly expanded to include new areas, including the suburbs of Phnom Penh. Its service is still expanding.

In 2008, the annual volume of treated water produced was a little over 85.5 million m^3, which was more than enough to meet the PPWSA's needs (Table 1). However, with increasing per capita water use from the lower base of earlier years, combined with an increase in its service area, the daily water production has to be increased to 300,000 m^3/day to meet anticipated demands by 2013. This increase can be met by increasing the capacity of the Chrouy Changwar plant by 65,000 m^3/day by 2013.

Progressive increases in the daily per capita water use in the city of Phnom Penh are shown in Figure 3.

Unaccounted for Water

A major problem in 1993 was substantial losses due to unaccounted for water (UFW) which was well over 70%. Reducing UFW within a short timeframe, a process involving several concurrent tasks, required a strict systems approach. No water supply can be kept at a reasonable and affordable level if the income from the customers does not cover the costs of providing a good service, especially when nearly three-quarters of the water supplied yielded no revenue whatsoever. Thus, a strict regime with several interrelated components had to be planned and implemented. The progressive reduction in UFW is shown in Figure 4.

No significant UFW reduction is possible without the cooperation and support of a dedicated, competent and motivated staff. The quality of staff in the PPWSA left much to

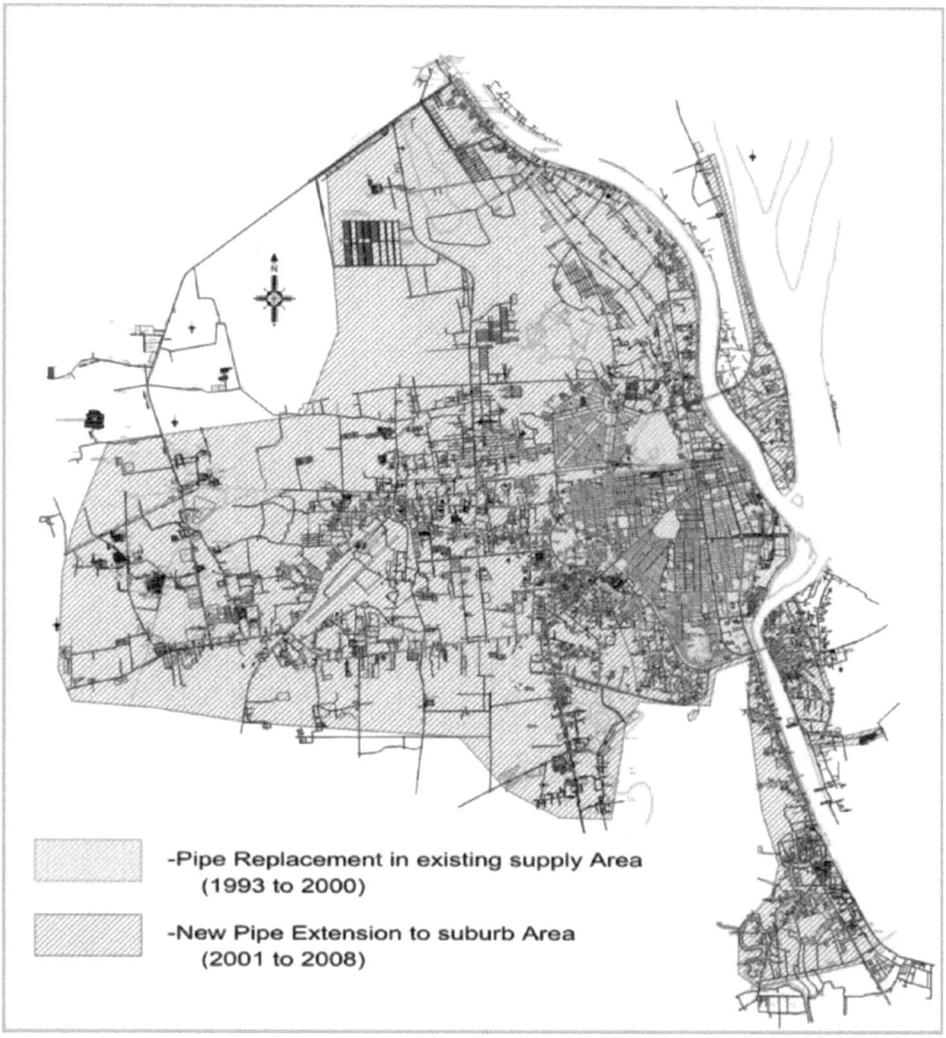

Figure 2. Expansion of area supplied with water, 1993–2008. *Source:* PPWSA records.

be desired during the early 1990s. Not only were the staff demoralized, but they had good reason to be demoralized, being, as they were, faced with poor governance, below subsistence pay, lack of discipline, an absence of any incentives, and pervasive corruption. Lethargy, poor working practices and a 'could not care less' attitude to consumers were the norm. Therefore, the work culture had to be radically changed by enforcing strict disciplines in a sensitive, fair and transparent manner. This was a difficult task since the rest of the public sector employees in Cambodia were in a similar situation and behaved in a very similar manner. In fact, most public sector companies still continue to behave in a somewhat similar way at present.

Table 1. Average annual water production in m^3/day

Year	Actual production (m^3)	Average daily production (m^3/day)	Max. daily production (m^3/day)
1993	19,586,000	53,660	No data
1994	25,483,000	69,816	No data
1995	31,651,000	86,715	No data
1996	34,486,000	94,482	No data
1997	39,184,171	107,354	116,646
1998	39,983,794	109,545	113,116
1999	40,622,401	111,294	117,837
2000	39,801,167	109,044	117,586
2001	37,763,547	103,462	103,978
2002	41,793,679	114,503	115,362
2003	46,871,146	128,414	144,405
2004	56,775,305	155,549	182,124
2005	61,857,961	169,474	190,367
2006	79,623,204	193,488	211,605
2007	79,400,031	217,534	240,349
2008	85,513,649	234,284	258,571

Source: PPWSA records

Changing institutional culture was not an easy task. It had to start with the senior officers who had to become the role models. During the 1980s and early 1990s, one of the perks of the job was that the employees of the PPWSA received a free supply of water. This practice was stopped. Staff members not only had to install meters but also had to pay their water bills in full, like any other citizen, and within the stipulated time period. Otherwise, they were treated the same way as those who were delinquent with their bills.

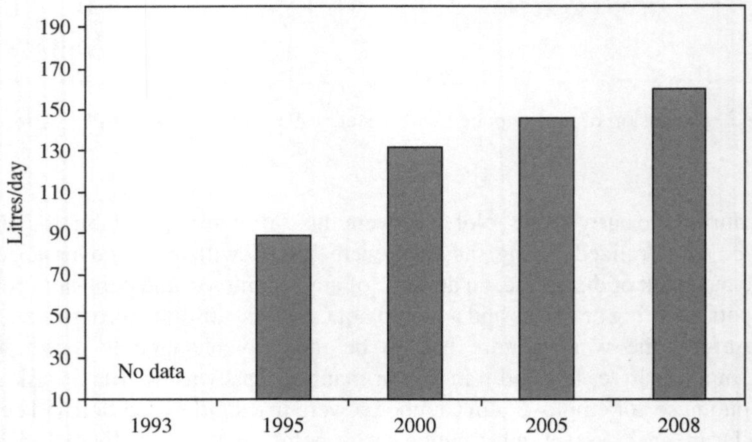

Figure 3. Per capita water use, 1993–2008. *Source:* PPWSA records.

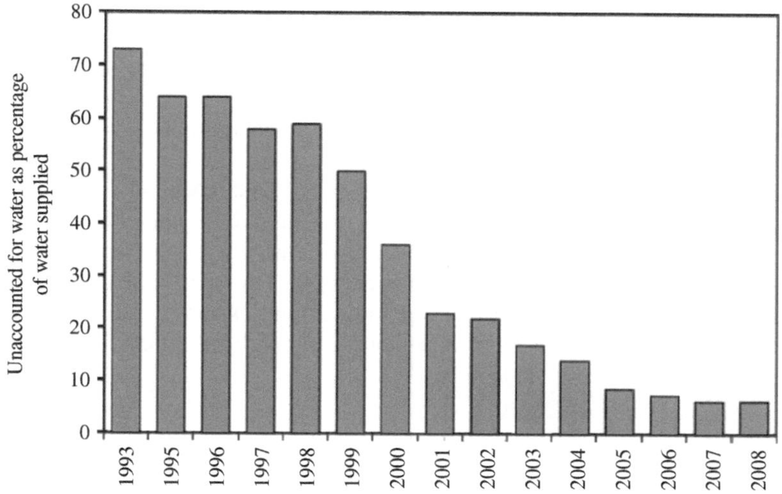

Figure 4. Unaccounted for water, Phnom Penh, 1993–2008. *Source:* PPWSA records.

Water Pricing and Cost Recovery

The uniform global experience is that if consumers have to receive a 24-hour, uninterrupted, and reliable water service, they have to pay for this service either directly or through taxes.

The government of Cambodia decided in the 1990s that water was an economic and social good, and thus a responsible, financially viable and socially sensitive tariff structure should be in place so that the PPWSA could have adequate income to be financially self-sufficient. The result would be that the institution would be run in a business-like way and the population should have good quality and affordable water supply.

To the credit of the Cambodia government, the PPWSA has been allowed to operate as an independent business-like institution without any political interference, with checks and balances as stipulated in the landmark Decree No. 52 of 1996. An English version of this decree can be found in Biswas & Tortajada (2009).

With its autonomous structure and good management, the PPWSA decided to maximize its income by:

- reducing UFW significantly so that much of the water produced can be sold to the consumers;
- fixing a tariff structure and implementing it fully with a social conscience;
- preparing and continually updating a roster of customers on a reliable basis;
- completely restructuring the billing system so that that the bills can be produced and delivered on time;
- improving the bill collection ratio with appropriate incentives, with disincentives of late or no payment; and
- increasing the annual profits of the PPWSA by becoming increasingly efficient on a progressive basis.

In order to achieve higher revenues and thus financial self-sufficiency, the PPWSA embarked upon a concurrent five-prong strategy. One of the most difficult components of this strategy to implement was to increase the tariff of water so that all costs could be recovered without generating any social and political unrest. This was done by ensuring that its customers first witnessed and appreciated a much better quality of reliable service before the tariffs were increased.

The increase in the tariff was very carefully planned. First, a socio-economic survey of the water supply situation was carried out for the city of Phnom Penh. This survey included collecting information on how much consumers were paying for water from private vendors, and what was likely to be their reaction if the supply by these vendors was replaced by the PPWSA. This survey showed the willingness and capability of consumers to pay a higher tariff than what they were being charged by the PPWSA provided they received a significantly improved service.

Table 2. Progressive tariff structure of PPWSA, in riels/m^3

Until 1983:	Free water for every household	
1984:	166 riels/m^3 (domestic and commercial)	
1993–June 1994:	166 riels/m^3 (domestic)	
	515 riels/m^3 (industrial)	
July 1994–May 1997:	250 riels/m^3 (domestic)	
	700 riels/m^3 (industrial)	
June 1997:	Block tariff rates were introduced. All connections were metered.	
	Domestic tariffs (m^3/month)	
	0–15:	300 riels
	16–30:	620 riels
	31–100:	940 riels
	>100:	1,260 riels
	Industrial tariffs (m^3/month)	
	0–100:	940 riels
	101–200:	1,260 riels
	201–500:	1,580 riels
	> 500:	1,900 riels
	Governmental institutions (m^3)	
	Flat rate:	940 riels
2001:	Domestic tariffs (m^3/month)	
	0–7:	550 riels
	8–15:	770 riels
	16–50:	1,010 riels
	>50:	1,270 riels
	Commercial tariffs (m^3/month)	
	0–100:	950 riels
	101–200:	1,150 riels
	201–500:	1,350 riels
	>500:	1,450 riels
	Governmental institutions (m^3)	
	Flat rate:	1,030 riels

Source: PPWSA records

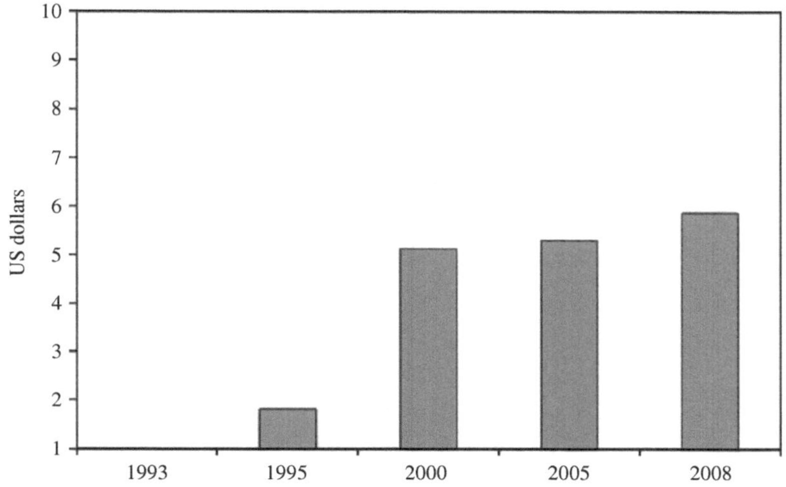

Figure 5. Average household water bill per month, US$, 1993–2008. *Source:* PPWSA records.

The tariff was calculated after considering the total expenses of the PPWSA, including operation and maintenance costs and the depreciation of all its assets. This process conforms to the National Water Policy which was approved later. Under this Policy, the utility is expected to recover all its operating costs with tariffs, as well as the depreciation of all its assets, except land which generally increase in value over time. The value of its assets has to be revised every five years. The second increase in tariff was introduced in 2001. The changes in the tariff structure from 1983 are shown in Table 2.

Having witnessed significant improvements in water delivery, including quality of water, the consumers paid for the vastly superior service, especially as they considered the increase to be very reasonable, and thus within their capacity to pay. This included the poor whose bills declined by 60–80% of what they had been paying to private water vendors. As a result, for all practical purposes, the private water vendors have disappeared from the PPWSA area.

By 1995, 24-hour uninterrupted service was available in Phnom Penh. This also meant that the average monthly household bills in the city increased significantly between 1995 and 2000, and thereafter have increased only incrementally (Figure 5).

While it was comparatively easy to convince the general public to pay a higher tariff, it was a different situation with respect to army installations, government offices and senior officials of the government and the army. They had not historically paid any water bill and not surprisingly, they wanted this free service to continue indefinitely. The army initially refused to have any meter installed to measure its actual water consumption, and refused to be billed for it. Nevertheless, in 1997, the Prime Minister publicly proclaimed that every person and institution had to pay their water bills promptly so that PPWSA could ensure good service delivery. The transition process was difficult.

Figure 6 shows the billing ratio in terms of quantity of water that was billed as a percentage ratio of total volume of water that was produced between 1993 and 2008. It shows that in 1993, less than 30% of water produced was billed. By 2005, with very

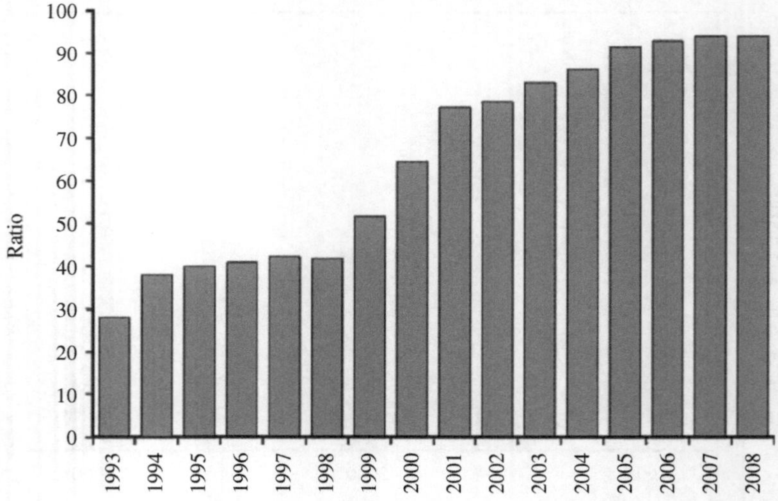

Figure 6. Billing ratio, quantity billed/total production in percentage. *Source:* PPWSA records.

significant reduction in UFW, over 90% of the volume produced was billed. The current situation is even better. At present, over 93% of water produced is billed.

Figure 7 shows bill collection ratios in terms of the amount collected and the amount billed. Both show bill collection is close to 100% from about 1999, more than double the rate in 1993 just six years earlier. This again is a most remarkable achievement.

Accurate Data on Water Consumers

For any functional and efficient water utility it is essential to have an accurate database of its current users. Without such an accurate database of the consumers, it is simply not

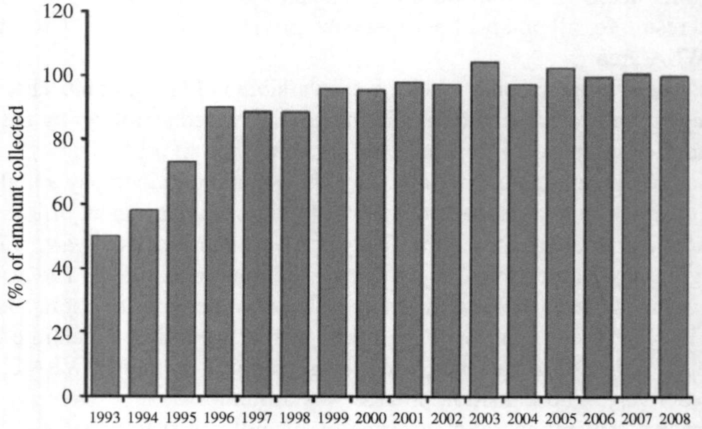

Figure 7. Bill collection ratio, amount collected/amount billed, in percentage, 1993–2008. *Source:* PPWSA.

possible to collect revenue from the water users. In addition, this database needs to be kept continually updated, which means keeping relevant information on the people who move into the city, people who move house but within the city, and also people who leave the city. Such a list is always dynamic, and needs to be continually updated to ensure correct and appropriate billing.

During the Khmer Rouge regime, all of the information on the PPWSA customers and other relevant information was destroyed. After the fall of the Khmer Rouge government, no accurate database was established as to who were the actual consumers of the water supplied by the PPWSA. In order to establish a proper database, in 1994 nearly 100 PPWSA members visited all of the houses in the city to record if they were receiving water or not. The process took one year to complete.

The results were revealing. The survey found that 12,980 households were ostensibly connected to the system, but were not. They were being billed for phantom water, and surprisingly many were paying their bills even though they never received any water. In contrast, there were 13,901 customers who were connected and receiving water, but were not in the list and thus were not being billed. They were receiving free water.

The French government gave a grant which enabled the PPWSA to establish a fully computerized up-to-date database, which became fully operational in 1996 (Chan, 2009). The computerized system was further improved in 2001 to handle all financial transactions and operations at the PPWSA. By 2003, a comprehensive financial management information system was in place which enabled the Authority to have immediate access to the latest, as well as historical financial data, and its revenue collection status in real time. An indirect benefit of this automated system has been that corruption and abuse of power were, for all practical purposes, eliminated. In addition, the Authority has continued to steadily improve its overall financial management practices.

Metering

In order to ensure that a transparent and fair system exists, it is essential that all of the connections should be measured in terms of their water consumption. Only after each connection is metered can the consumers be sent an accurate bill which directly reflects the amount of water they consumed during a specific period.

In 1993, only 3391 connections out of a total of 26,881 were metered. In other words, more than 87% of the connections received an estimated bill, which often had no linkage to the quantity of water that was actually consumed. With a policy decision to move to a system of all metered connections as soon as possible, the number of meters installed increased steadily, and the number of unmetered connection started to decline. By 2001, all the connections were metered. In addition, over time more accurate Class C meters replaced less reliable Class B meters, which further increased the credibility and income of the PPWSA.

Water for the Absolute Poor

The issue of providing clean water to the poor, or even to the rich, could not be considered prior to 1994. There were simply no reliable water services available, either for the rich or for the poor. Thus, the first task of the new management of the PPWSA was to ensure a reasonably reliable water supply system to the people irrespective of whether they are rich, middle class or absolutely poor. Hence, the primary focus and concern of the PPWSA at this initial stage was the rehabilitation of the water system in order to establish a supply system which could then be distributed to the people.

Once the supply was restored to a reasonable level, the issue arose about how water could be supplied to the poor in a way that the PPWSA could justify economically, and how the poor could afford the services they received. The first approach was to determine specific communities of poor people, who would then elect community representatives to whom the PPWSA would sell water, and who, in turn, were expected to resell water to their respective communities at a recommended tariff which the poor could afford. The installation of the community water pipes was initially funded mostly by certain NGOs. Thus, from 1995–2000, community representatives were supplied with water. By the end of 1998, there were 53 representatives of the poor communities, with an average total consumption of 15,500 m^3/month, representing slightly over one per cent of the total volume of water sales of the PPWSA.

The system did not perform as was initially expected for two reasons. First, the community representatives mostly sold water to the poor people at 10 to 20 times the subsidized rate the PPWSA sold water to them. Second, the PPWSA received very little income for the water sold through the community representatives. The main beneficiaries of the system turned out to be the community representatives themselves, who pocketed the difference between the two prices.

During the experimental period, the PPWSA also realized another important fact. Initially, it expected that the poor might steal water as a result of which its net UFW rate would remain high. Accordingly, meters were installed far from the poor households, which increased the installation costs. In 1999, the PPWSA realized that even if the poor stole water, which they did not, their individual consumption rate was so low that the resulting losses were unlikely to be significant.

In March 1998, the PPWSA received a soft loan from the World Bank, a part of which was explicitly earmarked for the provision of clean water for the poor.

A task force was created specifically for the implementation of this pro-poor scheme, which was also charged with the responsibility of processing and approving the applicants, and also implementing installment payments for the connection charges. This arrangement did not work very well, primarily because the necessary information was not properly disseminated to the poor households. Thus, in 1999 only 101 house connections to the poor could be made under this scheme. In contrast, for the city as a whole, 12,059 house connections were installed that year.

At present, thanks to a grant by the International Development Association (IDA), and an annual grant of 50,000 euros from the city of Paris, the PPWSA has a much more generous and affordable programme for the absolute poor. Poor households are entitled to receive subsidies of 30%, 50%, 70% or 100% of the connection fee, depending upon their financial conditions. These conditions are jointly evaluated by a committee of the PPWSA with direct help from the local communities. In addition, those households that consume a maximum of 7 m^3/month pay only 60% of the real cost of providing water. This new policy helped the poor households to save 130,000 to 380,000 riels each year (5000 riels = US$1.20).

As a result of these improvements in pro-poor policies, the number of poor households that has been connected to the system has steadily increased from 101 household connections in 1999 to 17,657 connections at present.

Financial Sustainability

On the basis of current evidence, the financial sustainability of the PPWSA can be summarized by one word: excellent. According to the latest statement from its auditors (Pricewaterhouse Coopers, 2009), its operating revenue for the 2008 financial year was 91,587.65 million riels, on which it made a profit of 30,577.58 million riels. After paying income tax of 6,141.35 million riels (20% of the profit) to the government of Cambodia, it made a net profit of 24,436.24 million riels. To our knowledge, it is the only publicly managed water utility in the developing world which has consistently increased its net profit since 1993, and also has paid consistently higher income taxes every year to the government. In fact, according to the audit report of 2008, the PPWSA actually made a loan of US$2 million (equivalent to about 8.4 billion riels) in 2005 to the *Electricité du Cambodge*, which was fully repaid in 2007. In 2008, the PPWSA gave another loan to the Pursat Water Supply Authority for 675.44 million riels to finance the construction of the main water supply network in the Kandiang district of Pursat Province. The unsecured loan is subject to 5% interest per year, to be reimbursed to the PPWSA in 120 monthly instalments, after the construction of the network was completed in 2009.

In 2009 the main components of the operating costs of the PPWSA were for electricity (44.97%) and chemicals needed for water treatment (10.73%), over which it has no control (Figure 8). The electricity costs are dictated by the *Electricité du Cambodge*, and the chemical costs are determined by world market prices. Salaries, wages and allowances for the staff account for another 32.16% of annual expenditure. While the Authority is becoming increasingly efficient in terms of its staff performance (for example, the number of accounts served per employee has dramatically improved between 1993 and 2008, as shown in Figure 9), the Authority must pay an attractive financial package to its staff in the future in order to recruit and retain a highly motivated staff.

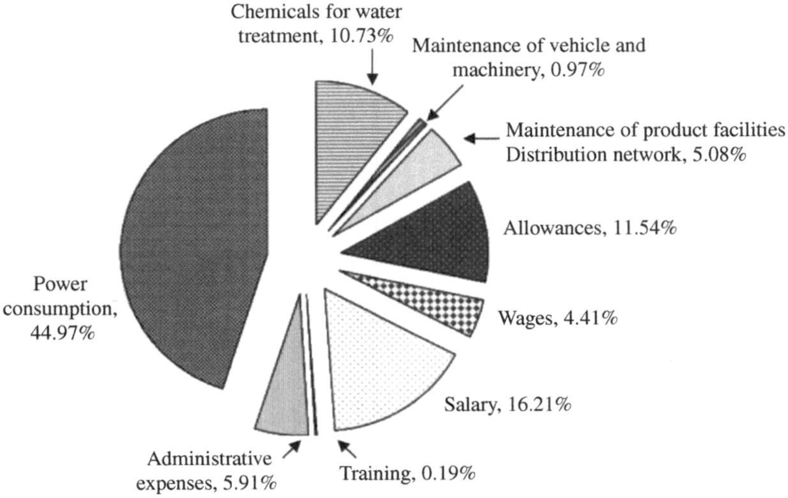

Figure 8. Distribution of annual operating expenses. *Source:* PPWSA records.

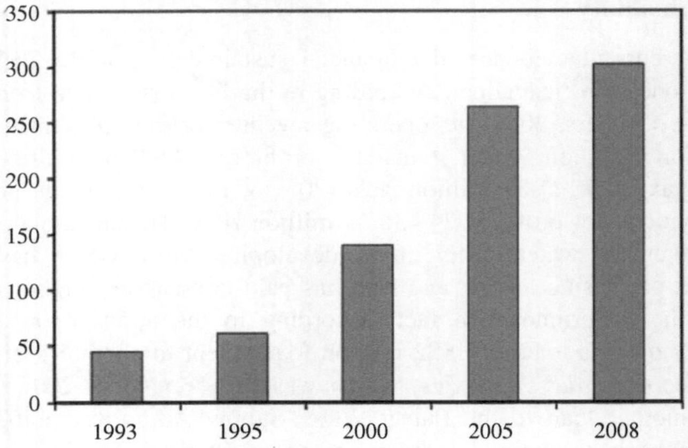

Figure 9. Number of accounts served per employee, 1993–2008. *Source:* PPWSA records.

With good efficiency gains, amounts billed and amounts collected became almost the same after 2000. This means that the net profit of the Authority has progressively increased every year since 1998 (Figure 10).

As the PPWSA increases its geographical area of coverage, more and more consumers join the system. Even with increasing per capita consumption, the Authority has maintained, and even improved upon its current level of performance. There is thus no reason why net profits should not continue to increase steadily, unless political interference reduces performance levels. At the current stage of development, it is highly unlikely that there would be political interference in its management, especially as it is the only public sector utility that has been consistently delivering good service to the population while concurrently increasing profit each year.

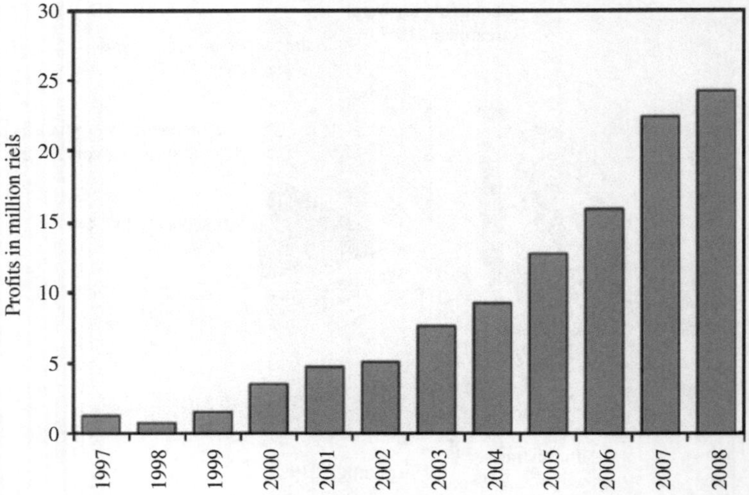

Figure 10. Net profits of PPWSA, 1997–2008. *Source:* PPWSA records.

PPWSA: An Overall Performance Assessment

In retrospect, if any independent and objective observer had been asked to make a realistic forecast on the future prospect of a near bankrupt, demoralized and a corrupt institution like the PPWSA in 1993, the most optimistic individual would probably have said "not very good." At best, the prognosis would have been that the Authority would be able to improve its performance incrementally, and would manage to provide water of uncertain quality to some of the inhabitants of Phnom Penh for a few hours per day. The most optimistic forecast would have been that by 2003, the PPWSA would become an average water utility in the region with a mediocre, or perhaps even less than mediocre, performance.

Instead, the actual achievements of the PPWSA have confounded everyone, including its most ardent supporters and the donors. In only one decade, the institution has turned around 180 degrees, with a completely different mindset, a team spirit that is seldom observed in a utility (water, gas or electricity) in any developing country, and a 'can do' attitude which is very refreshing to observe. From 1994, and every year since that time, the Authority has made significant progress in all of the technical, managerial and financial performance indicators. It has continuously expanded its network, improved its management and operating efficiency, become financially self-sufficient, and progressively increased its net annual profit after paying appropriate income taxes to the Cambodian government—as one would expect from any private sector institution.

Within the 15-year timeframe, 1993–2008, the PPWSA increased its annual water production by 437%, distribution network by 557%, pressure of the system by 1260%, and customer base by 662%. During the same period, it reduced UFW losses from 72% of treated water produced in 1993 to only 6.19% at present. By judicious use of incentives and sanctions for its staff with transparent policies that were consistently implemented, and a strong and determined focus on capacity building for all its staff, the number of accounts served per employee increased by 671% during the same period.

During the same period, the number of metered connections has gone up by nearly 5255%, and the number of accounts handled per employee has increased by 671%. Even though people are now paying for the water they actually consume, per capita daily water consumption has nearly doubled, which must have a positive impact on the health and well-being of Phnom Penh residents.

Concluding Remarks

In terms of good governance, the experience of the PPWSA is a salutary example for the urban centres of the developing world. Many reasons are given by water utility managers, political leaders, and members of the water profession as to why it has not been possible to provide clean, drinkable water to the urban centres of the developing world. Among these has been the physical scarcity of water, lack of availability of investment funds, inability of the poor to pay for water, lack of expertise, and among many other reasons. In our view, all of these are only excuses to hide the real and fundamental reason for this shortcoming—poor leadership and governance of the urban water sector nearly all over the developing world. Phnom Penh has very clearly shown how it can be achieved under the most difficult circumstances, and in less than ten years.

IMPROVING WATER POLICY AND GOVERNANCE

Unless the urban water governance practices are improved significantly, as in the case of Phnom Penh, universal access to clean drinking water will remain an unachievable dream, even if hundreds of billions of dollars with no strings attached are made available to this sector each year.

Acknowledgements

The funding for this independent evaluation came from the Institute of Water Policy, Lee Kuan Yew School of Public Policy, National University of Singapore.

References

ADB (1996) Report and Recommendation of the President to the Board of Directors on a Proposed Loan to the Kingdom of Cambodia for the Phnom Penh Water Supply and Drainage Project. Asian Development Bank, Manila.

Biswas, A. & Tortajada, C. (2009) Water Supply of Phnom Penh: A Most Remarkable Transformation. Research Report. Third World Centre for Water Management, Atizapan, Mexico.

Chan, E. S. (2009) Bringing safe water to Phnom Penh's city, Water, *Water Resources Development*, 25(4), pp. 597–609.

Pricewaterhouse Coopers (2009) Phnom Penh Water Supply Authority Financial Statements for the year ended 31 December 2008, Phnom Penh.

Tokyo Engineering Consultants & Nihon Suido Consultants (1993a) *The Study on Phnom Penh Water Supply System in the Kingdom of Cambodia*, Final Report, Vol. 1, Summary, Japan: JICA.

Tokyo Engineering Consultants & Nihon Suido Consultant (1993b) *The Study on Phnom Penh Water Supply System in the Kingdom of Cambodia*, Final Report, 3 Volumes, Japan: JICA.

Service Quality and Performance Measurement: Evidence from the Indian Water Sector

SURENDER KUMAR* & SHUNSUKE MANAGI**

*Department of Policy Studies, TERI University, New Delhi, India; **Graduate School of Environmental Studies, Tohoku University, Aoba-Ku, Sendai, Japan and Institute for Global Environmental Strategies, Hayama, Kanagawa, Japan

ABSTRACT *The conventional measures of benchmarking focus mainly on the water produced or water delivered, and ignore the service quality, and as a result the 'low-cost and low-quality' utilities are rated as efficient units. Benchmarking must credit utilities for improvements in service delivery. This study measures the performance of 20 urban water utilities using data from an Asian Development Bank survey of Indian water utilities in 2005. It applies data envelopment analysis to measure the performance of utilities. The results reveal that incorporation of a quality dimension into the analysis significantly increases the average performance of utilities. The difference between conventional quantity-based measures and quality-adjusted estimates implies that there are significant opportunity costs of maintaining the quality of services in water delivery.*

Introduction

By 2025, 50% of Indians will reside in urban areas and will face acute water problems (Planning Commission, 2002). The reasons for prevailing water scarcity in many cities can be attributed not only to source limitation but also to poor efficiency in distribution networks. Inefficiency in the delivery of water services translates into negative health impact and social unrest. Benchmarking is an essential tool for improving the performance of water utilities. The conventional measures of benchmarking concentrate only on the water produced or water delivered and ignore the service quality. As a result the 'low-cost and low-quality' utilities are rated as efficient units. It is estimated that about 12.2 million people die every year due to diseases directly related to drinking contaminated water in developing countries (Lin & Berg, 2008).

This study measures the performance of 20 urban water utilities using data from the Asian Development Bank (ADB) survey of Indian water utilities in 2005 (ADB, 2007). The study recognizes that the service delivery requires additional resources. Hours of supply and pass rate of chlorine tests are considered as indicators of quality in water service delivery. Data envelopment analysis (DEA) is applied for measuring the performance of the utilities under varying returns to scale. Measuring performance under varying returns to scale helps in designing the optimal scale of operation and has implications for water pricing. It was found that the performance of utilities changes significantly when conventional quantity-based measures are compared with the quality-adjusted measures. Performance of an average water utility is 0.53 and 0.62 (where the value close to 1 is efficient) when measured using quantity-based and quality-adjusted measures respectively, implying that the opportunity cost of maintaining the service delivery is about 9% of the potential water delivered.

The remainder of the paper is organized as follows: The next section presents an analytical sketch of some typical facts of urban water supply systems in India. The section following reviews the relevant literature and discusses the empirical strategy followed in the present study. Next is a section presenting the empirical application that measures the performance of Indian urban water utilities; it discusses the methodology, data and results. The study closes with some concluding remarks.

Urban Water Supply in India

After six decades of independence, about 64–74% of the rural population and about 91–93% of the urban population still draws its drinking water from protected sources (Ghosh & Goldar, 2007). Here, sources considered protected are piped water and hand pumps. If the wide disparity in the coverage across states is factored in, the picture is very dismal. The erratic nature of the supply and degrading water quality further worsen the situation.

The World Health Organization (WHO) issues guidelines for drinking water quality to mitigate the health risks posed by contaminated drinking water (WHO, 2007). The first requirement is a sufficient water quantity. The per capita water consumption in 20 cities in India is shown in Figure 1.[1] The average litres per capita per day (lpcd) for the sample cities is about 128 lpcd compared to the standard of 140 lpcd fixed by the National Drinking Water Mission in the late 1980s. About 40% of the sample cities attain the norm. In the sample cities, lpcd varies between 72 for Bhopal to 203 in Jamshedpur.

In term of hours of water supply, among the 20 urban localities, Chandigarh fares well. The water supply in cities such as Rajkot, Visakhapatnam and Indore has been very poor, ranging between 25 minutes to one hour per day (Figure 2). The average hours of water supply among cities has been 4.3 hours. If we compare these figures with those of some other Asian cities, we find that most of the Indian cities analysed here perform poorly. For example, cities such as Singapore, Hong Kong, Seoul and Kuala Lumpur are supplied water 24 hours everyday (Gupta *et al.*, 2006).

Sporadic water supply coupled with insufficient pressure and unpredictable services impose both financial and health costs. Based on a household survey, Misra & Goldar (2008) find that the coping cost[2] for an average household in Delhi is Rs 10 per kilolitre (kl) and Rs 187 per month in addition to the average water and sewerage bill of Rs 140 per month (Rs = rupee). This coping cost is in addition to the bill paid by the consumers to Delhi Jal Board (DJB). The operating cost of the DJB is Rs 6 per kilolitre. This shows that the consumers in Delhi have been effectively paying Rs 16 per kilolitre due to poor services in water delivery.

The second requirement is that the water should not be contaminated. Given the WHO guidelines, tests are conducted for positive rate of chlorine. Figure 3 shows the variability in the quality of water delivered by the water utilities in sample cities. Monitoring of water

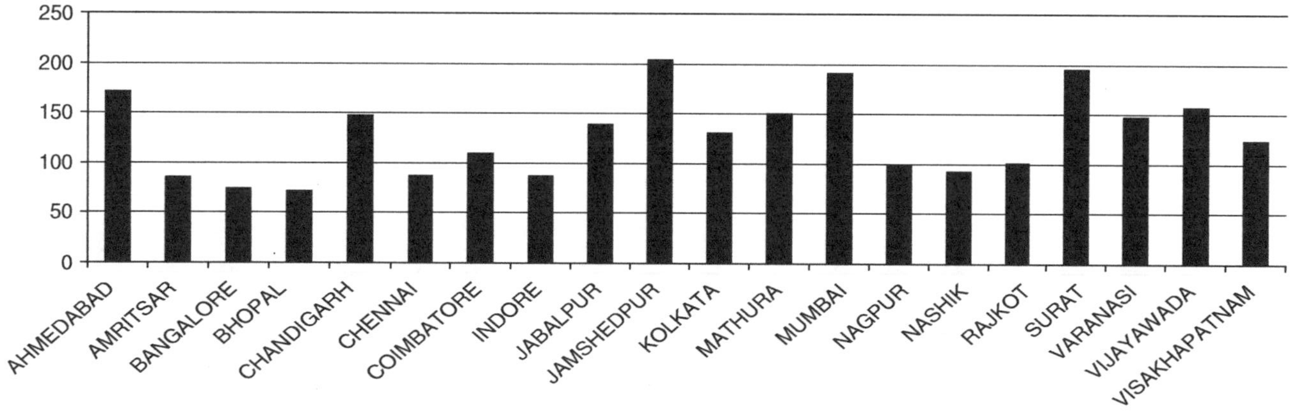

Figure 1. Per capita water consumption (lpcd) *Source:* ADB, 2007.

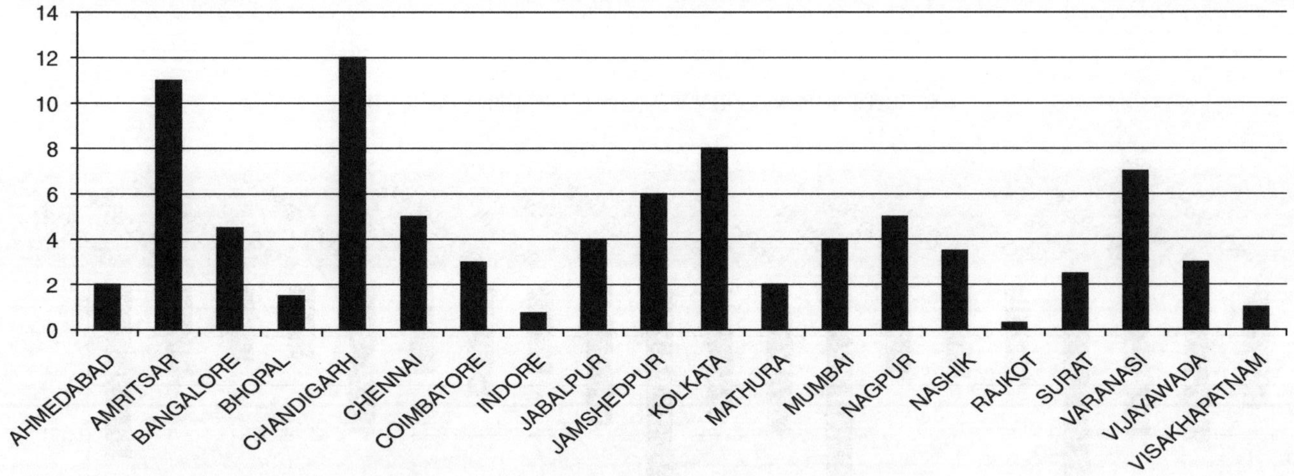

Figure 2. Water availability (hours/day) *Source:* ADB, 2007.

quality in Indian cities is haphazard. McKenzie and Ray find that in Kolkata about 87% of water reservoirs serving residential buildings and 63% of taps had high levels of faecal contamination. About 90 million workdays per year are lost in terms of treatment costs and loss of production due to waterborne diseases in India—an economic cost of about Rs 6 billion (McKenzie & Ray, 2009).

One of the main reasons behind water shortages in urban India is high water losses in the distribution network. The water produced but not reaching consumers can be considered a standard indicator of the inefficiencies of water utilities. In the cities analysed, the average water loss, measured as 'unaccounted for water' (UFW), is about 32% of water produced. In some cities, such as Bhopal, Amritsar, Nagpur and Nasik, the UFW is more than 50% of the produced water (Figure 4). This indicates a substantial scope for improvement.

Background and Empirical Strategy

Usually two different categories of performance benchmarking techniques are employed to measure the efficiency of water utilities. One is the single-measure gap analysis approach, which generally refers to the activity of studying the differences between standards and the delivery of those standards, such as using the number of workers per connection, or the number of connections per 1000 families, for assessing the performance of water utilities. Generally, multilateral lending agencies such as the World Bank and the ADB utilize these measures for assessing the performance of public utilities. The second technique takes into account all of the multiple inputs used and outputs produced by the utilities and is known as the 'total factor productivity' (TFP) approach. The former cannot be a substitute for the latter, though it is the simplest form of performance evaluation and benchmarking. For instance, the high ratio of staff per connection in India may not indicate the inefficiency of the water utilities. It could be due to cheaper labour substituting for other inputs rather than overstaffing. Therefore, it is important to use the multiple inputs–outputs frontier model to evaluate utility efficiency. The TFP approach recognizes the complex nature of interactions between multiple inputs and multiple outputs.

The TFP studies apply either one of two techniques for measuring performance of public utilities: DEA, and stochastic frontier analysis (SFA). There are two essential differences between the SFA and DEA. The SFA attempts to distinguish the effects of data noise from the effects of inefficiency. The DEA technique requires neither specification of the error term nor the distributional form of the error term; the major disadvantage of this approach is that it does not accommodate the effects of data noise.

DEA finds a production frontier consisting of the most relatively technically efficient utilities in the sample. This process generates the technical efficiency measures for each unit in the sample by comparing observed values (the particular data point) to optimal values (best performing value) of outputs and inputs. A score of 1.0 represents the best performing unit in the sample, and a score less than 1.0 implies that the unit is not performing as well as its efficient peers. A rather interesting implication of the DEA score is how much more output could have been produced if the given unit could somehow emulate the production process of the efficient one, that is, the utility operating at the frontier of production technology.

In the last three decades, many studies in developed and developing countries have examined the performance of water utilities using either DEA or SFA, or both of the techniques. Empirical applications have been driven by the different concerns. Most of the studies have been conducted with the objective of examining the impact of ownership

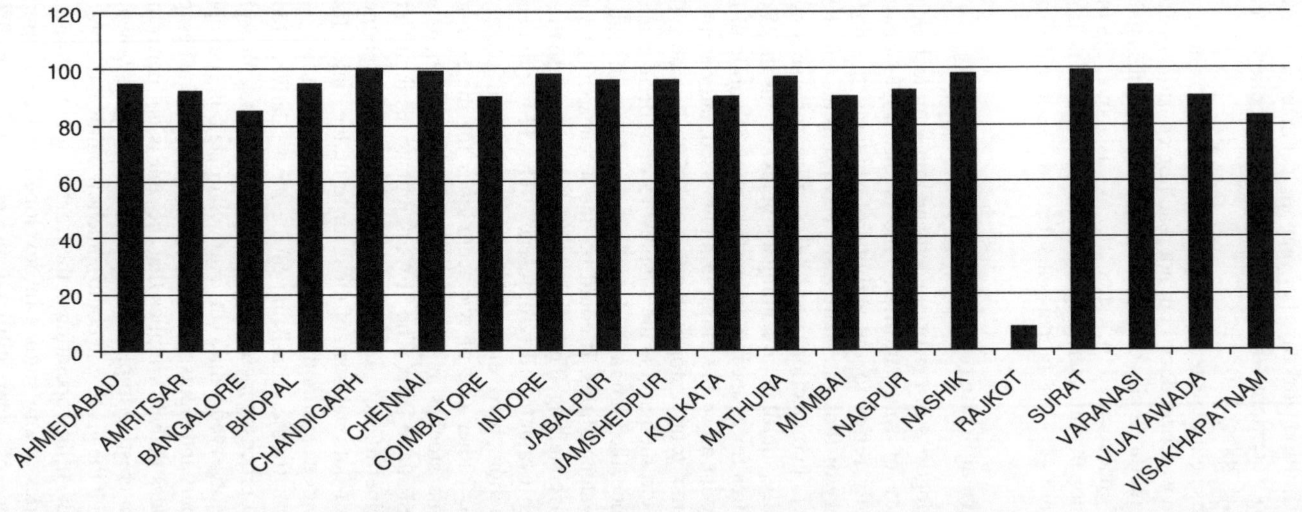

Figure 3. Positive rate of chlorine tests (%) *Source:* ADB, 2007.

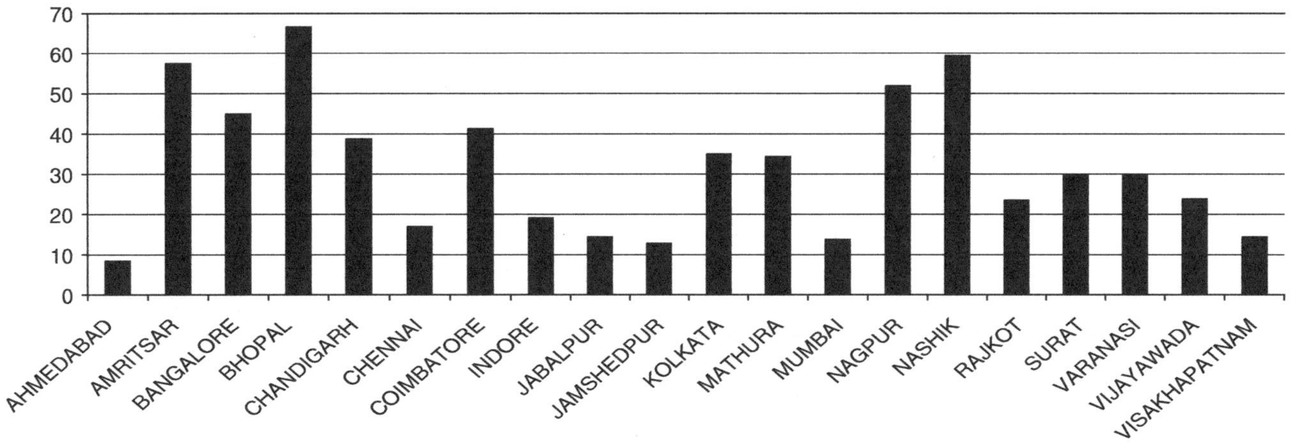

Figure 4. Unaccounted for water (% age of produced water) *Source:* ADB, 2007.

(public versus private) on the performance of water utilities. Studies have also been conducted examining the effects of public regulations on utility performance, the extent of scale economies, or the influence of operating environments on utility performance. Some of the recent studies include Cubbin (2005), Kirkpatrick *et al.* (2006), and Garcıa-Valinas & Muniz (2007). A common feature of all these studies is that they ignore quality as a relevant dimension of the services provided by the water utilities.

Saal & Parker (2000, 2001) provide, perhaps, the first studies that started to underscore the importance of quality in measuring the performance of water utilities. These studies incorporate the quality dimension by using quality-adjusted indices of output in measuring the performance of the water and sewerage industry in England and Wales. The results suggest the existence of quality-driven scope economies. Consequently, some very recent studies have specifically focused on assessing the quality-adjusted performance of water utilities. These include the studies by Lin (2005), Saal & Parker (2006), Saal *et al.* (2007), Picazo-Tadeo (2008) and Lin & Berg (2008). Saal & Parker (2006) and Saal *et al.* (2007) observe that quantity-based measures fail to show any impact of privatization on productivity improvement, but find substantial impact on the productivity of the industry in terms of quality-adjusted estimates. Picazo-Tadeo *et al.* (2008) estimate the performance of Spanish water utilities and consider the UFW as an indicator of quality. Lin (2005) and Lin and Berg (2008) consider quality dimensions as additional outputs, and find that including quality variables not only changes benchmarking results but also alters the ranking of utilities.

Despite the existing inefficiency concerns in the urban water supply situation in India, there is a surprising dearth of literature assessing the performance of utilities using TFP approaches. Most studies use partial productivity measurement methods to examine efficiency, for example, Singhal & Johri (2002); World Bank (2006); ADB (2007). The Water and Sanitation Program (World Bank, 2006) and the Asian Development Bank (ADB, 2007) developed some performance indicators to measure the efficiency of water supply systems in some select urban localities. The WSP collected performance data for 13 utilities covering 23 cities and towns, and the ADB collected data for 20 cities across India. The indicators chosen by both studies are similar, such as investment, financing, billing and collection, quality, costs and staffing, network, metering, UFW, production/consumption, and coverage, among others. These studies applied the single measure gap analysis method to evaluate the cities in terms of their ability to supply water, and are incapable of reflecting the true picture of the performance of the utilities.

Gupta *et al.* (2006) is perhaps the first study in India estimating the technical efficiency of 27 urban water utilities using the TFP approach. The study applies the output-oriented DEA approach for estimating relative efficiency in the provision of urban water utilities for the year 2004–2005, and uses data from the City Development Plans (CDPs) prepared by the municipalities for getting financial assistance under the Jawaharlal Nehru National Urban Renewal Mission (JNNURM). It considers total water produced as a function of the revenue expenditure and production capacity of a water plant, and ignores the quality concerns in the service delivery.

Commonly, in the performance measurement, total water produced or water delivered is considered a function of available resources. The efforts of utilities made towards providing better services are ignored. When resources are directed at providing better quality services, rather than producing only billed water, the output/input ratio of the utility is lower and the efficiency of the utility appears lower. It is the presence of an unobserved constraint on the utility that causes the output efficiency measure to identify the output combination as technically suboptimal. Thus, the conventional quantity-based measures of performance penalize utilities that produce higher quality services. Accordingly, the 'low-cost low-quality' utilities tend to display, on equal terms, a greater score for efficiency.

The non-parametric DEA approach was used to provide a framework for a more complete representation of production technology of Indian urban water utilities. This approach incorporates both the delivered water and quality of services for measuring 'how far' each utility's output vector is from the best practice frontier for a given input vector. It seeks to enhance delivered water and water quality measured in terms of more hours of supply and the pass rate of chlorine tests. This approach recognizes that service quality is a desirable output. That is, some productive resources have to be given up in order to provide quality services.

Methodology

The technology of a water utility can be specified either by production, cost or profit function. The output distance function generalizes the production technology of a multi-output multi-input utility. Assume that a water utility employs a vector of inputs $x = (x_1, \ldots\ldots, x_N) \in R_+^N$ such as operation and maintenance (O&M) expenditure and capital stock to produce a vector of outputs $y = (y_1, \ldots., y_M) \in R_+^M$ such as accounted for water (AFW), hours of supply, and quality of water. The production technology of the water utility is defined as:

$$T = \{(x, y) : x \, \text{can produce} \, y\} \tag{1}$$

and the output distance function is defined as,

$$D_o(x, y) = \inf [\theta : \{x, y/\theta\} \in T] \tag{2}$$

Equation (2) characterizes the output possibility set by the maximum equiproportional expansion of all outputs consistent with the technology set (1). The output distance function is non-decreasing and positively linearly homogeneous. It is also convex in outputs and decreasing in inputs. The output distance function takes a value, which is less than or equal to 1.0 if the output vector is an element of the feasible production set. Furthermore, the distance function will take a value of 1.0 if the output vector is located on the outer boundary of the production possibility set.

DEA was applied to compute the output distance functions. The DEA involves the application of linear programming to construct a piecewise linear envelopment frontier over the data points such that all observed points lie on or below the frontier. Technical efficiency for each utility is computed as:

$$\begin{aligned}
(D_y(x^0, y^{k'}))^{-1} &= \max \theta \\
&s.t. \\
\sum_{k=1}^{K} z_k y_m^k &\geq \theta y_m^{k'} \quad m = 1, \ldots., M \\
\sum_{k=1}^{K} z_k x_n^k &\leq x_n^0 \quad n = 1, \ldots., N \\
z_k &\geq 0 \quad k = 1, \ldots\ldots, K
\end{aligned} \tag{3}$$

where M is the number of outputs produced, N is the number of inputs employed, and K is the number of water utilities. ($z_1, \ldots.., z_k$) are variables which show the intensity with which each utility is used in order to construct the frontier of the production possibilities set. In the formulation of equation (3), by restricting $z_k \geq 0$, we impose the condition of constant returns to scale (CRS). In the DEA, various kinds of returns to scale on reference technology can be imposed by changing the restrictions on intensity variables. For example if $\sum_{k=1}^{K} z_k \leq 1$, then the technology satisfies the non-increasing returns to scale (NRS), and if the sum of intensity variable is restricted to exactly equal to 1.0, we can model variable returns to scale (VRS). See Managi *et al.* (2004) for a conceptual description of DEA.

Technical inefficiency under CRS is the product of scale inefficiency and pure technical inefficiency. A water utility is operating under the condition of optimal returns to scale if the output-oriented technical efficiency of a water utility is equal under CRS and VRS scenarios, otherwise it is scale inefficient. Deviation from scale efficiency is essentially deviation from CRS. If the technical efficiency under CRS is not equal to the efficiency under VRS, but the efficiency score under VRS and NRS are equal, then the utility is operating under decreasing returns to scale (DRS). Finally, if the efficiency scores are equal under CRS and NRS, but less than VRS, then the utility is operating in the range of increasing returns to scale (IRS) (Grosskopf, 1986).

Data and Results

The principal source of data for this study is the ADB's *2007 Benchmarking and Data Book of Water Utilities in India* (ADB, 2007). ADB provides information from water utilities in 20 JNNURM cities and is based largely on 2005–2006 data. The cities covered in the ADB study are: Ahmadabad, Amritsar, Bangalore, Bhopal, Chandigarh, Chennai, Coimbatore, Indore, Jabalpur, Jamshedpur, Kolkata, Mathura, Mumbai, Nagpur, Nasik, Rajkot, Surat, Varanasi, Vijayawada, and Visakhapatnam.

As mentioned, water utilities are considered as production units. Selecting the output and input variables to be included in a DEA model is always a difficult decision. Three outputs are considered in this study: the AFW, hours of supply, and water quality. The AFW is considered the quantitative dimension of output of the utility, and hours of supply and the pass rate of chlorine tests are considered as the indicator of the quality of service. Given the sporadic supply and near to the ground service quality of water delivered by the water utilities in India, hours of supply and pass rate of chlorine tests are adequate measures of the quality of service of water utilities (Picazo-Tadeo *et al.*, 2008). The choice of input vector is based on the availability of data. We consider two inputs: O&M expenditure incurred in the year as a proxy for variable input, and the sum of capital expenditure incurred in the last five years as a proxy for capital stock. Though it is a very strong assumption, given the data limitations, there is no better choice available. The other input could have been the storage capacity of the utilities, but the utilization of storage capacity itself is linked to the efficiency of the concerned utility and depends on the resources available. Thus, for estimating the technical efficiency, it is assumed that the utilities are producing AFW, hours of supply and water quality using two inputs—O&M expenditure and capital stock. The descriptive statistics of the variables used are provided in Table 1.

Estimates of quantity-based and quality-adjusted technical efficiency under CRS, VRS and NRS have been computed by solving equation (3) for each utility in the sample. Recall that an output-oriented measure of technical efficiency is being used, therefore,

Table 1. Descriptive statistics of the variables used in the study (2005-2006)

Variable	Measurement unit	Mean	Standard deviation	Maximum	Minimum
Accounted for water	Million cubic meter	138.04	215.27	1,008.76	9.12
O&M expenditure	Million rupees	750.08	1,129.22	4,284.06	28.12
Capital stock	Million rupees	2,126.40	3,979.66	17,342.50	20.60
Water availability	Hours	4.30	3.20	12.00	0.33
Pass rate of chlorine tests	Per cent	89.35	19.70	100.00	8.00
Per capita per day Water supply	Litres	128.20	41.33	203.00	72.00
Unaccounted for water	Per cent of total water produced	31.85	17.37	66.50	8.30

Source: ADB, 2007.

efficiency scores greater than 1.0 imply that the utility has potential to increase its output for the given level of inputs. For the sake of convention, the results are presented by taking their reciprocals. The efficiency is estimated in three scenarios: considering only the AFW; AFW and hours of supply; and AFW, hours of supply and pass rate of chlorine tests. The results are shown in Figures 5–7, respectively, for pure technical efficiency (i.e., VRS), scale efficiency and optimal technical efficiency (i.e., CRS).

Technical Efficiency Estimates

Figure 5 shows the quantity-based estimates of technical efficiency of water utilities. Ahmadabad, Bhopal, Jabalpur and Varanasi are shown to be the most efficient sample utilities. These utilities are not only technically efficient, but they are also operating at the optimal scale. In Ahmadabad and Varanasi, the O&M expenditure per cubic meter of water is lowest, which are Rs 1.52 and Rs 1.90, respectively. Similarly in the case of Jabalpur it was found that the ratio of accounted for to total produced water is next to highest, the first being Ahmadabad. In the case of Bhopal, the utility becomes the outlier due to low capital stock per unit of cubic water billed. Bangalore and Indore are the utilities found to be least efficient, as these utilities are incurring the highest per cubic meter O&M expenditure (ADB, 2007). The mean values of technical efficiency under CRS and VRS and scale efficiency are 0.53, 0.67 and 0.78, respectively. An average Indian urban water utility has the potential to increase the accounted for water by 47% for the given level of resources. About 22% of the output of billed water can be increased by operating at the optimal scale, and the remaining inefficiency can be attributed to other managerial factors.

Figures 6 and 7 present the quality-adjusted technical efficiency estimates. When the estimates of technical efficiency are adjusted for hours of supply and pass rate of chlorine test, there are six utilities on the frontier and operating at optimal scale. Amritsar and Mathura are the two additional utilities operating at the frontier of the production possibilities set. The mean values of technical efficiency under CRS, VRS and scale efficiency are 0.62, 0.78 and 0.76, respectively, when adjusted for hours of supply only, and are 0.62, 0.94 and 0.66, respectively, when adjusted for both hours of supply and pass rate of chlorine tests. For an average Indian water utility the potential to increase the

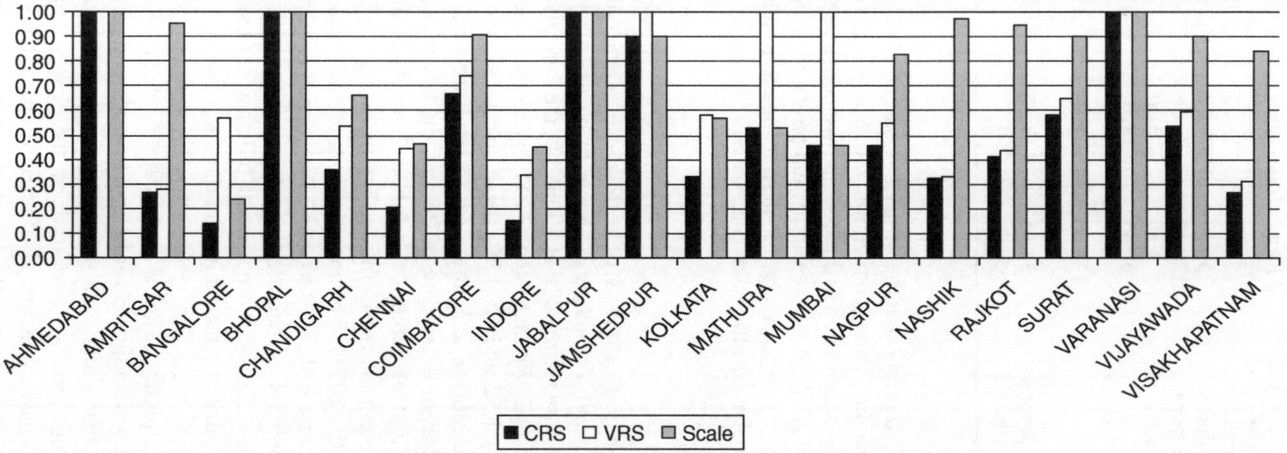

Figure 5. Performance of urban water utilities (only accounted for water).

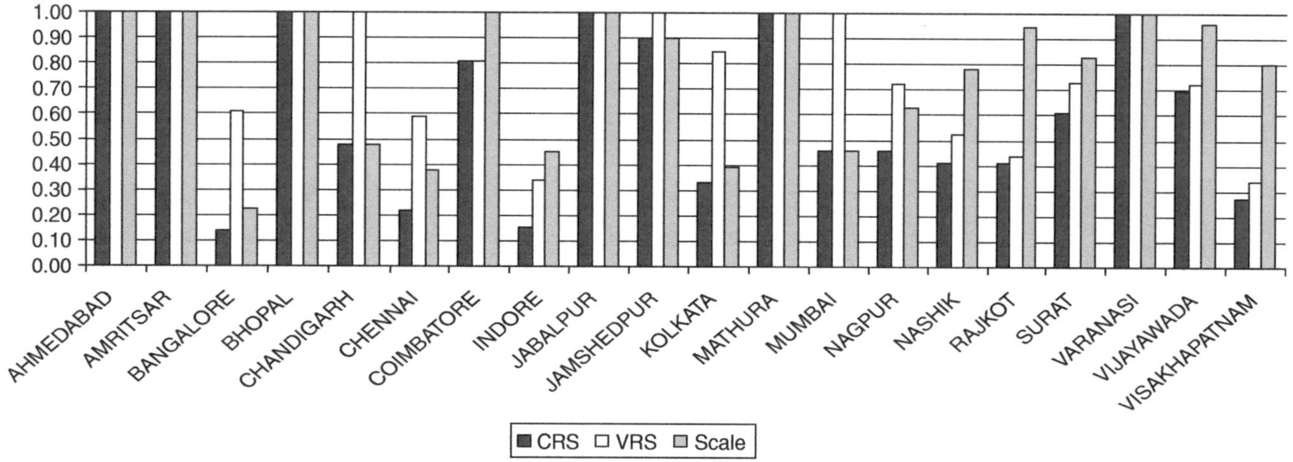

Figure 6. Performance of urban water utilities (accounted for water and hours of supply).

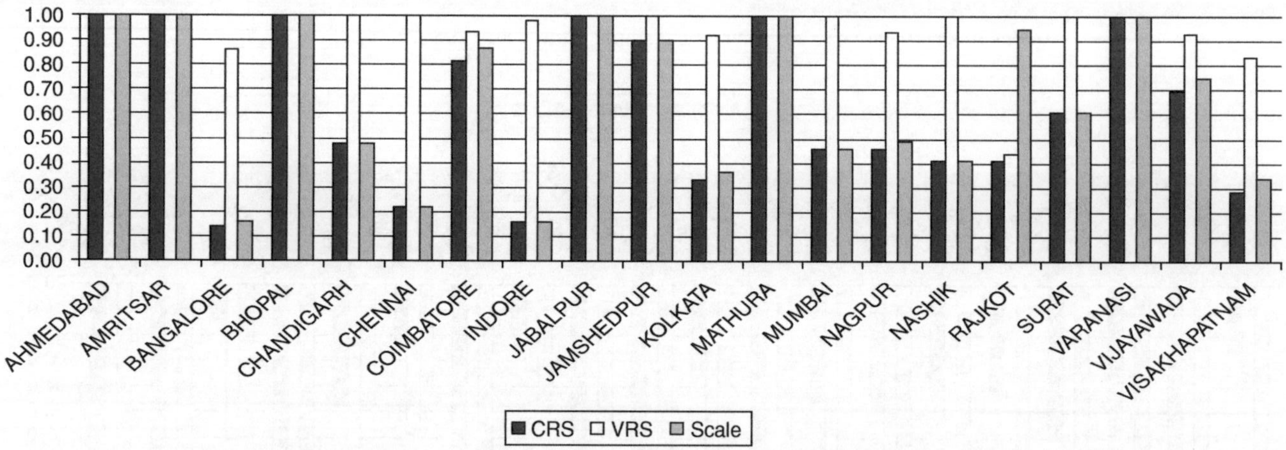

Figure 7. Performance of urban water utilities (accounted for water, hours of supply, and water quality).

quality-adjusted quantity of water is about 38 million cubic metres. About 34% of the inefficiency is due to the choice of an inappropriate scale of operation, and the remainder can be attributed to other managerial factors.

To obtain an economic interpretation of the estimates of technical efficiency adjusting for quality changes, Coimbatore was randomly chosen. For this utility, the conventional quantity-based measure of technical efficiency is 0.67 under CRS. This implies that, given the level of its resources, it is producing about two-thirds of its potential output. This is irrespective of the level of quality achieved at the projection on the efficient frontier that determines this potential output. Conversely, forcing this utility to maintain hours of supply and pass rate of chlorine tests at the observed levels increases the score of technical efficiency under CRS to 0.81. This implies that potential output is reduced as some productive resources must be unavoidably directed towards maintaining the quality. The opportunity cost of maintaining quality for this utility reaches 14% of its potential output and under VRS, the opportunity cost of maintaining quality is about 19% of its potential output.

One of the major concerns of the study is to find the answer to the question: "Does service quality matter in measuring the performance of water utilities?" The results obtained here show that it does, as is revealed by the difference between conventional quantity-based and quality-adjusted measures of technical efficiency. To confirm the answer, a t-test was conducted for equality of means for evaluating the statistical significance of this difference, and a non-parametric Wilcoxon signed rank test for equality of distributions. In addition to these tests, the Spearman rank correlation test was conducted to determine whether the sample water utilities rank differently according to their conventional and quality-adjusted measures of performance. Test results are provided in Table 2, where conventional quantity-based measures are the reference point for comparison.

The results of the t-test under CRS reveal that the null hypothesis about the equality of means of technical efficiency between the quantity-based and quality-adjusted estimates has not been rejected. There is no statistically significant difference between the means of conventional and the quality-adjusted estimates. But the null hypothesis is rejected when the means are compared using the estimates obtained under VRS. Similarly, the Wilcoxon test is examined. The null hypothesis of equality of distributions of quantity-based and quality-adjusted estimates is rejected under both the CRS and the VRS at a confidence level of 99% (p-values are negligible). However, on the basis of the Spearman rank correlation coefficient, the null hypothesis of independence of estimates from each other

Table 2. Tests for the difference between quality-based and quality-adjusted (for both, hours of supply and positive rate of chlorine tests) measures of technical efficiency.

	t test[a] (t-stat)	Wilcoxon test[b] (z-stat)	Spearman test[c] (Spearman ρ)
Quantity-based versus quality-adjusted (under CRS)	−0.911 (0.37)[d]	−3.109 (0.0019)	0.818 (0.00)
Quantity-based versus quality-adjusted (under VRS)	−4.018 (0.0003)	−3.370 (0.0008)	0.407 (0.075)

Notes: [a]The null hypothesis is the equality of means. [b]The null hypothesis is that both distributions are same. [c]The null hypothesis is that both variables are independent.[d]Values in parentheses are p-values.

is also rejected at 99 (negligible p-value) and 90 (0.075 p-value) per cent levels of significance. That is, the ranking of utilities under the two different scenarios is statistically not different from each other. The Spearman rank correlation coefficient is 0.82 and 0.41 under CRS and VRS, respectively. These results imply that though the mean and distribution of efficiency scores in both evaluations is statistically different, the water utilities cannot be ranked differently when ordered according to either quantity-based estimates or quality-adjusted estimates of technical efficiency. This finding implies that managerial abilities continue to play an important role in determining the performance of Indian urban water utilities.

Scale Economies

Technical efficiency is the product of scale efficiency and pure technical efficiency. Figures 5–7 show the operating scale of different water utilities under different scenarios. The overall average figures of scale efficiency reveal that the utilities are not utilizing their resources optimally. These water utilities can improve their performance by changing the level of their operation. The scale efficiency results indicate that four and seven out of 20 utilities are operating at the optimal scale according to quantity-based and quality-adjusted estimates, respectively (Table 3). The remaining utilities are operating under either increasing or decreasing returns to scale. About 25% of sample units (Coimbatore, Mathura, Nasik, Rajkot, and Vijayawada) are operating under IRS, and all of the remaining water utilities are operating under DRS according to conventional measures of technical efficiency. The utilities that are operating under IRS are generally the utilities of relatively small cities. However, we find that only the Rajkot utility is operating under IRS when the results are adjusted for quality variables. These results have implication for urban domestic water pricing. In the public utility pricing literature, it is assumed that the utilities are operating under IRS and the marginal cost-pricing rule that ensures economic efficiency is not applied since the full cost is not recovered. These results support the idea that to improve efficiency in the operation of water utilities, the water should be priced according to the marginal cost of water supply. This finding corroborates that of Whittington (2003) which also observes that many South Asian cities are facing the situation of decreasing returns to scale in operation as they have to incur high costs to bring additional water into cities.

Policy Implications

Performance measurement establishes how effectively and efficiently a public utility delivers the required service. It accounts for both the quantitative and qualitative aspects of an agency's functioning, and it relates policy options and their outcomes.

This study finds that without any consideration for quality in water service delivery, an average Indian urban water utility has the potential to increase the accounted for water by 47%, out of which about 22% could be realized by operating at the optimal scale, and to realize the remaining potential there is a need to focus on managerial factors. However, when quality is taken into account, the potential to increase the accounted for water is about 38%. Out of this 38%, about 34% potential could be realized by operating at the optimal scale. Moreover, we find that most Indian water utilities are operating under DRS. These findings imply that: (a) about two-fifths more water could be delivered to consumers

Table 3. Returns to scale

Returns to scale	City
Quantity based estimates of technical efficiency	
CRS[a]	Ahmadabad, Bhopal, Jabalpur, Varanasi
DRS[b]	Amritsar, Bangalore, Chandigarh, Chennai, Indore, Jamshedpur, Kolkata, Mumbai, Nagpur, Surat, Visakhapatnam
IRS[c]	Coimbatore, Mathura, Nasik, Rajkot, Vijayawada
Estimates of technical efficiency adjusted for hours of supply	
CRS	Ahmadabad, Amritsar, Bhopal, Coimbatore, Jabalpur, Mathura, Varanasi
DRS	Bangalore, Chandigarh, Chennai, Indore, Jamshedpur, Kolkata, Mumbai, Nagpur, Nasik, Surat, Vijayawada, Visakhapatnam
IRS	Rajkot
Estimates of technical efficiency adjusted for hours of supply and pass rate of chlorine tests	
CRS	Ahmadabad, Amritsar, Bhopal, Jabalpur, Mathura, Varanasi
DRS	Bangalore, Chandigarh, Chennai, Coimbatore, Indore, Jamshedpur, Kolkata, Mumbai, Nagpur, Nasik, Surat, Vijayawada, Visakhapatnam
IRS	Rajkot

Notes: [a]CRS: constant returns to scale; [b]DRS: decreasing returns to scale; [c]IRS: increasing returns to scale.

if the existing resources were utilized optimally; (b) Indian water utilities are not operating at the optimal scale; the marginal cost-pricing rule could be followed to improve efficiency in the delivery of water services; (c) delivery of quality services is not without costs and for the Indian water utilities this cost is about 9% of accounted for water; and (d) managerial abilities play an important role in determining the performance of Indian urban water utilities.

The estimates of utility performance obtained using conventional and adjusted measures could be of certain use for policy makers and utility management. Given the context of a developing country where contaminated water causes significant negative health outcomes, the estimates obtained by this study might have some interest from a social perspective. Achieving a better understanding of the cost of maintaining quality, and the nature of trade-offs between quantity and quality, should help improve the management of utilities. The results might also be of some use to local public representatives, as there are talks of reforming urban civic amenities in India either through privatization of water utilities or some combination of public and private management of these services. The results can also be useful for regulators as there is talk of introducing independent regulators in the water sector at the state level. Maharashtra has already introduced a regulator for the water sector.[3] The quality-adjusted performance measure can provide useful information to regulators about the Q (quality) factor in the quality-dependent price cap regulation (i.e., CPI-X+Q, where CPI is the consumer price index and X is the productivity offset) (Lin & Berg, 2008). Furthermore, given the fact that water is a scarce natural resource and there is scarcity of resources at the level of local bodies, identifying utilities that are employing better management practices has a certain interest for society as a whole.

Conclusions

Water is a scarce natural resource and inefficiencies in the water sector translate into negative health impacts and social unrest (Hattingh & Claassen, 2008.). The estimates of benchmarking are important for evaluating the performance of reforms, comparing the relative performance of different utilities, informing citizens who might exert pressure in order to improve performance, and helping regulatory bodies to provide incentive frameworks for improving the performance of the utilities (Lin, 2005).

The conventional measures of benchmarking concentrate only on the water produced or water delivered and ignore the service quality, and as a result the 'low-cost and low-quality' utilities are rated as efficient units. Benchmarking must credit utilities for improvements in service delivery. This study measures the performance of 20 urban water utilities using data from an ADB survey of Indian water utilities in 2005. The study recognizes that the service delivery requires additional resources. It uses DEA for measuring the technical efficiency of water utilities. It also measures the performance of utilities under different returns to scale to help in designing the optimal scale of operation and water pricing. The study finds that the performance of utilities varies when conventional quantity-based measures are compared with the quality-adjusted measures. It finds that the performance of an average water utility is only 0.53 and 0.62 when measured using quantity-based and quality-adjusted measures. The results imply that quality matters and the opportunity cost of maintaining the service delivery is about 9% of the potential water delivered. The study also finds that most Indian water utilities are operating under DRS implying that the water should be priced according to the marginal cost pricing rule. Future research needs to address productivity change over time in terms of efficiency change, technological change, and quality change, and relating the changes with utility-specific characteristics and policy variables so that policy lessons can be learned. As more data become available, and additional studies are conducted, a better understanding of the role of quality in the water sector will be possible.

Notes

1. ADB (2007) computes per capita per day water consumption using the following formula: Per capita per day water consumption = {[Total annual domestic consumption (M3) × 1000/365]/number of people served}, where annual water consumption is defined as the difference between annual water production and unaccounted for water.
2. The coping strategies adopted by consumers to manage water shortages, irregularity of supply, low pressure, water quality problems, and the like, include investments and recurring expenses in borewells, booster pumps, water storage, purchase of water (bottled water/vendor supply), use of filters, aqua-guards and boiling of water. Coping costs also include opportunity cost of time spent on water collection (Misra & Goldar, 2008)
3. Maharashtra appointed the Maharashtra Water Resource Regulatory Authority (MWRRA) in 2005. The mandate for the MWRRA includes for overseeing the relationship between the service provider and water user entities and also within a water user entity, in terms of determination, enforcement and dispute resolution of entitlements and fixation of water charges. Independence of water regulation in India is in its early stages. Legislation to introduce water regulators is in process in at least three other states — Arunachal Pradesh, Gujarat and Delhi (Dubash, 2008)

References

Asian Development Bank (ADB) (2007) *2007 Benchmarking and Data Book of Water Utilities in India* (Manila: Asian Development Bank).

Cubbin, J. (2005) Efficiency in the water industry, *Utilities Policy*, 12, pp. 289–293.

Dubash, N. (2008) Independent regulatory agencies: a theoretical review with reference to electricity and water in India, *Economic and Political Weekly*, 4 October, pp. 43–54.

Garcıa-Valinas, M. A. & Muniz, M. A. (2007) Is DEA useful in the regulation of water utilities? a dynamic efficiency evaluation (a dynamic efficiency evaluation of water utilities), *Applied Economics*, 39, pp. 245–252.

Ghosh, M. & Goldar, B. N. (2007) *Safe Water, Safe Water Disposal and Health Outcomes* (Delhi: Institute of Economic Growth).

Grosskopf, S. (1986) The role of the reference technology in measuring productive efficiency, *The Economic Journal*, 96, pp. 499–513.

Gupta, S., Kumar, S. & Sarangi, G. K. (2006) *Measuring the Performance of Water Service Providers in Urban India: Implications for Managing Water Utilities*, NIUA WP 06-08 (New Delhi: National Institute of Urban Affairs).

Hattingh, J. & Claassen, M. (2008) Securing water quality for life, *International Journal of Water Resources Development*, 24(3), pp. 401–415.

Kirkpatrick, C., Parker, D. & Zhang, Y. F. (2006) An empirical analysis of state and private sector provision of water services in Africa, *World Bank Economic Review*, 20(1), pp. 143–163.

Lin, C. (2005) Service quality and prospects for benchmarking: evidence from the Peru water sector, *Utilities Policy*, 13, pp. 230–239.

Lin, C. & Berg, S. V. (2008) Incorporating service quality into yardstick regulation: an application to the Peru water sector, *Review of Industrial Organization*, 32, pp. 53–57.

Managi, S., Opaluch, J. J., Jin, D. & Grigalunas, T. A. (2004) Technological change and depletion in offshore oil and gas, *Journal of Environmental Economics and Management*, 47(2), pp. 388–409.

McKenzie, D. & Ray, I. (2009) Urban water supply in India: status, reform options and possible lessons, *Water Policy*, 11, pp. 442–460.

Misra, S. & Goldar, B. N. (2008) Likely impact of reforming water supply and sewerage services in Delhi, *Economic and Political Weekly*, October 11, pp. 57–66.

Picazo-Tadeo, A. J., Saez-Fernandez, F. J. & Gonzalez-Gomez, F. (2008) Does service quality matter in measuring the performance of water utilities? *Utilities Policy*, 16, pp. 30–38.

Planning Commission (2002) *India Assessment 2002: Water Supply and Sanitation*, New Delhi: Planning Commission, Government India.

Saal, D. S. & Parker, D. (2000) The impact of privatization and regulation on the water and sewerage industry in England and Wales: a translog cost function model, *Managerial and Decision Economics*, 21, pp. 253–268.

Saal, D. S. & Parker, D. (2001) Productivity and price performance in the privatized water and sewerage companies of England and Wales, *Journal of Regulatory Economics*, 20, pp. 61–90.

Saal, D. S. & Parker, D. (2006) Assessing the performance of water operations in the English and Welsh water industry: a lesson in the implications of inappropriately assuming a common frontier, in: T. Coelli & D. Lawrence (Eds) *Performance Measurement and Regulation of Network Industries*, pp. 297–328 (Cheltenham: Edward Elgar).

Saal, D. S., Parker, D. & Weyman-Jones, T. (2007) Determining the contribution of technical change, efficiency change and scale change to productivity growth in the privatized English and Welsh water and sewerage industry: 1985–2000, *Journal of Productivity Analysis*, 28, pp. 127–139.

Singhal, S. & Johri, R. (2002) An approach directed towards developing a performance measurement system for urban local bodies for better management of water supply services in India, *Journal of Environmental Studies and Policy*, 5(1), pp. 53–62.

Whittington, D. (2003) Municipal water pricing and tariff design: a reform agenda for South Asia, *Water Policy*, 5, pp. 61–76.

World Bank (2006) *India—Water Supply and Sanitation: Bridging the Gap between Infrastructure and Service* (New York, NY: World Bank).

World Health Organization (WHO) (2007) *Guidelines for Drinking-water Quality*, 3rd edn (Geneva: World Health Organization).

Municipal Water Supply Management in Bangkok: Achievements and Lessons

MUKAND S. BABEL*, ALDRIN A. RIVAS* & SEETHARAM KALLIDAIKURICHI**

*Water Engineering and Management, School of Engineering and Technology, Asian Institute of Technology, Pathumthani, Thailand; **Institute of Water Policy, Lee Kuan Yew School of Public Policy, National University of Singapore, Singapore

ABSTRACT *Providing reliable and safe drinking water through a piped water system in Bangkok has enormous challenges owing to the city's physiographic and demographic characteristics. However, with its commitment to good corporate governance principles and practices, the implementation of required measures by the Metropolitan Waterworks Authority (MWA) has resulted in significant improvements in water supply services in Bangkok. At present, almost 100% of the population are provided with reliable and safe tap water through the state-run utility which has achieved a remarkable financial performance. The article presents and discusses the factors responsible for the achievements by the MWA to draw lessons for public policy and good practices which can be replicated in other urban areas, especially in developing countries. Moreover, areas for further improvement are suggested for better performance and quality of water services.*

Introduction

Urban centres, especially in the developing world, are confronted with tremendous challenges in providing adequate and safe water supply to its constituents (Biswas, 2006; Cohen, 2006). These are compounded with inadequate investments in water supply infrastructure, limits on water availability, and high rates of population growth (Van der Bruggen *et al.*, 2009), not to mention the lack of systematic land use planning and governance issues, among others. While many water utilities in cities in the Asian region seem unable to surmount the difficulties, a number of utilities appear to have risen above the challenges. Bangkok is one such case which has recently made significant improvements.

Bangkok, the capital of Thailand, is situated on the low flat plain of the Chao Phraya River which extends to the Gulf of Thailand. The Bangkok Metropolis is located at 13.45° N 100.28° E, and has a mean elevation of 2.31 m above mean sea level. With a total area of 1,569 km^2, the city is divided into 50 districts and 154 sub-districts (BMA & UNEP, 2004). Though Bangkok is highly urbanized with approximately 60% of the land area built-up, it also comprises a substantial agricultural area, about 29% of the metropolitan total, located largely in the periphery. Areas utilized for aquaculture activities cover around 5% while water bodies account for about 1%.

Bangkok has a large population. The registered population in 2008 was approximately 5,710,000[1], which is an increase of less than 1% from a decade ago. For greater Bangkok, which includes the adjacent provinces of Samut Prakan, Nonthaburi, Nakhon Pathom, Pathumthani, and Samut Sakhon, the registered population in 2008 was around 10.2 million with an area of approximately 7,760 km^2.

Bangkok has a monsoon type of climate which can be classified into three main seasons: rainy (May–October), cool (November–January), and hot (February–April). Rainfall in Bangkok is modest, with an average annual rainfall of approximately 1500 mm from 1971 to 2000. This amount, however, does not significantly influence water supply management issues as Bangkok meets its water needs from the Chao Phraya River and underlying aquifers which are fed or recharged from other areas receiving significantly higher rainfall. On the other hand, the occurrence and intensity of rainfall is of a significant concern. About 25% of rainfall occurs in September, whereas rainfall is minimal or almost negligible in the dry months from November to April. This presents challenges for discharging excessive runoff during the rainy season during which high rainfall intensity is common.

Domestic and industrial water supplies in Bangkok are provided by a combination of surface water and groundwater, the latter being mainly used in the outskirts of Bangkok. The Metropolitan Waterworks Authority (MWA, the Agency) manages the water supply that covers Bangkok and two adjacent provinces, Nonthaburi and Samut Prakan, a total area of 3,195 km^2. In 2006, the MWA supplied piped water of about 4.66 million m^3/day (equivalent to 91% of total demand) to residential, industrial and commercial sectors using surface water withdrawn from the Chao Phraya and Mae Klong rivers. The water is treated by conventional processes before distribution. The remaining 9% of the water demand (about 0.5 million m^3/day) was met by abstracting from deep wells (Polprasert, 2007), largely by industries, which prefer and are allowed by the Department of Groundwater Resources (DGR) to use groundwater for their operations or for manufacturing.

In assessing the management practices of the MWA, the paper discusses factors relevant to the achievements of the utility in terms of water supply, quality, and demand management. It then discusses and analyses governance aspects in the areas of water pricing, financial management, human resource management, service delivery, and corporate governance to draw lessons (both positive and negative) that can be learned from the MWA's experience. Suggestions are made as well to further improve the performance and water supply services.

Historical Overview of Water Supply Provision in Bangkok

Bangkok became the capital of Thailand shortly after General Taksin decided to transfer the capital from Ayutthaya to a site nearer to the sea which would facilitate foreign trade, ensure the procurement of arms, and make defence and withdrawal easier in case of renewed Burmese attacks. The new capital was established in Thonburi on the west bank of the Chao Phraya River. After the death of General Taksin, General Chakri became the first king of the Chakri Dynasty as King Rama I (King PhraBuddhayodfah) and ruled the kingdom from 1782 to 1809. His first action was to transfer the royal capital across the river from Thonburi to Bangkok to increase defense against attacks from Burma as Thailand, formerly known as

Siam, still had conflicts with Burma. The river formed natural defence moats on the west and the south sides of the city. At that time the water from the rivers, canals, and from rainfall was used as potable water and for other domestic purposes (AIT, 1996). To adjust to the lifestyle of the people, many canals were constructed and connected to the Chao Phraya River especially during the reigns of King Rama I to King Rama IV. Due to the numerous canals in Bangkok, it was later referred to as the "Venice of the East".

In the reign of Rama V (King Chulalongkorn, 1869–1910), the King expressed his concern about the quality of potable water. Having travelled to Europe a few times, Rama V mandated the supply of safe and clean water to the people of Bangkok as in European countries (MWA, 1998). On 13 July 1909, by Royal command, the waterworks were started with the following tasks: (1) construction of water storage at Khlong Bang Luang–Chiang Rak in Pathumthani province; (2) construction of canal to convey water from the water storage to Khlong Sam Sen; and (3) construction of a water treatment plant at Sam Sen.

During the reign of Rama VI (King Vajiravudh, 1910–1925), the production and distribution systems came into operation when the Bangkok Waterworks Authority inaugurated its services on 14 November 1914. The Bangkok Waterworks Authority was then under the Department of Sanitation, Ministry of Municipality until 1939, then under the Bangkok Municipality until 1952, and finally under the Municipal Public Works Department until the establishment of the MWA in 1967.

Developments related to water supply provision in neighbouring areas followed the developments in Bangkok. On the west side of the Chao Phraya River, the Thonburi Waterworks Authority initially obtained treated water to supply to its area of responsibility from the Bangkok Waterworks Authority via the distribution pipes along the King Rama I Bridge. When bombs damaged the bridge during World War II disrupting the transfer of treated water from Bangkok to Thonburi, the Thonburi Municipality started its own waterworks operations by drilling deep wells in 1953 and supplied water to the people in the area until the MWA took over the operations in 1968 following the merging of the utilities.

Similarly, in the south, the Samut Prakan Waterworks Authority commenced its services on 20 June 1936 under the control of Samut Prakan Municipality until 1969. In Nonthaburi province, located northwest of Bangkok, the Nonthaburi Waterworks Authority started its operations on 15 January 1960 under the supervision of the Municipal Public Works Department, Ministry of the Interior until the MWA took over the operations on 14 December 1967.

MWA: Mandates and Institutional Arrangement

The Metropolitan Waterworks Authority was established as a state enterprise through the Metropolitan Waterworks Authority Act 1967, enacted on 16 August 1967 by King Rama IX (King Bhumibol Adulyadej). The Act consolidated four preceding waterworks, namely, the Krungthep (Bangkok) Waterworks Authority, the Thonburi Waterworks Authority, the Nonthaburi Waterworks Authority, and the Samut Prakan Waterworks Authority. The Act mandated the MWA to survey, procure, and acquire raw water, as well as to produce, supply, and distribute treated water to the public in Bangkok, Nonthaburi, and Samut Prakan. The MWA is committed to its mission of providing cost-effective water supply services with high standards, modern technology, and competent professionals to ensure customer satisfaction.

One of the most important characteristics of the MWA as a state enterprise is the autonomy of its management and operations. The Board of Directors is given full authority to make decisions to manage everything within the state enterprise (MWA, 2000); only major projects and those related to budget and investments need subsequent approval from the Ministry of Interior (MoI), the regulating agency of the MWA. By virtue of the MWA Act, the MWA is authorized to set its water tariff without requiring the approval from a higher entity. This, therefore, provides the MWA with the freedom to formulate policies and programmes as well as to implement measures towards carrying out its mission, all of which has led to its successes over the years.

The Board of Directors of the MWA is responsible to set policy guidelines and provide general supervision over the MWA. Members of the board are appointed by the cabinet for a four-year term. Past practices in the appointment of members of the board indicate that representatives from relevant government agencies such as MoI (with the permanent secretary or the deputy as the chair of the board), the Ministry of Finance (MoF), and the Royal Irrigation Department (RID), among others, were usually appointed members of the Board. This has been changed, however, with the 2007 amendment of the Standard Qualification for Board of Directors and Staff of State Enterprises Act (RTG, 2007) in which members of the board are now selected by the cabinet from a list of eligible individuals, prepared by the MoF, who can be directors for state enterprises without regard to their institutional affiliations.

This development may be considered positive in the sense that the cabinet may be able to select without constraint the most qualified and appropriate individuals. On the other hand, the board may be composed of members with no representatives from relevant agencies such as MoI, MoF or RID, and that this may result in missing the perspectives or insights of these agencies in the development of the MWA policies and programmes. Although some agencies such as MoI and MoF monitor the MWA in some way, subsequent objections along the process could result in delaying the implementation of relevant programmes. Government agencies that are relevant to the functions of the MWA include:

- Ministry of Finance: monitoring the performance of the MWA through the State Enterprise Performance Appraisal process, review of the financial aspects of MWA operations and measures/projects;
- National Economic and Social Development Board (NESDB): review of development plans and large projects for further approval by the cabinet;
- Royal Irrigation Department: allocation of raw water;
- State Enterprise Policy Office: oversees financial and accounting policies of state enterprises;
- provincial government administrations, such as the Bangkok Metropolitan Administration (BMA), and the provincial governments of Samut Prakan and Nonthaburi: coordination for infrastructure activities and provision of services and support;
- Metropolitan Electricity Authority (MEA): provision of electricity/power for Bangkok;
- Ministry of Public Health: monitoring of tap water quality; and
- Royal Thai Police: assistance in facilitating civil works.

At the top of the management of the MWA is the MWA Governor—also a member of the Board—who is the chief executive assisted by six deputy governors in the areas of administration, services, engineering and construction, production and transmission, finance, and planning and development. In the past, the governor was selected exclusively from the senior officials of the MWA. However, in 2001, the MWA made it a policy to open the position, thereby giving opportunities to qualified individuals from other organizations to apply. This removed the constraints in selecting the most qualified individuals to lead the MWA. Along with procedures for selecting the members of the Board, this policy reflects the utility's commitment to the continuous improvement of its management strategies and approaches by providing a window for a broader perspective which may be brought in by individuals from outside of the MWA.

The selection of a suitable and strong leader is important for the success of an organization as has been demonstrated in the MWA itself. A significant contribution to the foundation of the current MWA management environment as a state enterprise was laid two decades ago. The governor who headed the MWA between 1984 and 1987 introduced the concept of operating and managing the MWA not as a typical government or semi-government entity, but like a business organization committed to public service. He gained the support of the national government for its projects, activities and management measures, some of which were not usually supported by the national government in the past. Employee bonuses were initiated, which was perceived to contribute to improving the performance and efficiency of MWA staff. Three revisions and adjustments to the water tariff made during 1984–87 contributed to the improved financial status of the utility.

Some changes in the structure of MWA organization occurred from time to time, and these contributed to positive outcomes. These changes, which are associated mainly with the number of staff, workload, workflow, and the responsibilities of each office, were made in response to emerging needs and developments as well as to improve the efficiency of personnel and organization. For example, changes made in 1997 were in accordance with changes in strategies due to the financial crisis that hit the country. The current structure of the organization is expected to support and respond to the MWA's strategic and management plans.

Approach and Achievements

Water Supply Management

As mandated, the MWA is responsible for water supply provision in Bangkok metropolis and the adjacent provinces of Nonthaburi (in the northwest) and Samut Prakan (in the south) with a total area of 3,195 km^2 and a registered population of 7.91 million in 2008. The MWA's water supply system is divided into East and West Bank systems with reference to the Chao Phraya River (Dhungana, 2002) and further subdivided into 15 areas. These areas are served correspondingly by 15 branch offices that handle water provision, customer service, pipe and valve repairs, meter replacement, meter recording, bills collection and other related services.

Until several years ago, the MWA utilized groundwater to supplement surface water sources. Groundwater use by the MWA peaked in 1981 at approximately 0.5 million m^3/day, whereas abstraction from the registered wells of private entities in Bangkok and neighbouring provinces (Nonthaburi, Samut Prakan, Pathumthani, Samut Sakhon and Ayutthaya) was approximately 0.9 million m^3/day in 1981 and continued to increase until 1997 when it reached about 2.2 million m^3/day (Babel *et al.*, 2007). The actual groundwater withdrawal could be much higher considering the presence of unregistered wells and inadequate monitoring and enforcement (Srivardhana, 1994). The excessive use of groundwater in Bangkok and neighbouring provinces brought negative impacts such as groundwater depletion, land subsidence, and water quality deterioration (Babel *et al.*, 2007). Thus, the national government intervened and issued in 1983 a Cabinet Resolution on Remedial Measures for the Mitigation of the Groundwater Crisis and Land Subsidence in Bangkok (Babel *et al.*, 2007) in order to reduce pumpage in the critical zones, which include the areas serviced by the MWA and adjacent areas. The MWA was directed to terminate the use of groundwater. The MWA did so without substantial effects on its operations and finances through a gradual reduction in and phasing out the use of groundwater wells, while at the same time acquiring new surface water sources and increasing the capacity of its water treatment plants. The construction of the West Bank Raw Water Canal, which brings water from the Mae Klong River to the Mahasawat WTP and subsequently to the MWA distribution system, commenced in 1993 and was completed in 2002 (MWA, 2003). In 2005, the MWA ceased abstraction and use of groundwater completely, but groundwater wells are currently kept as back-up sources for emergency needs.

At present, the sources of water for the MWA service area are the Chao Phraya and Mae Klong rivers, from which the MWA is allowed to divert 60 and 45 m^3/s, respectively. Being endowed with relatively abundant water resources, the MWA has been able to continuously expand its production capacities and service coverage as guided by its master and expansion plans. The production volume has increased by approximately 24% from 3.88 million m^3/day in 1999 to 4.82 million m^3/day in 2008. The MWA currently has four water treatment plants (WTPs) (Bangkhen, Mahasawat, Thonburi, and Samsen) with a total production capacity of 5.52 million m^3/day. Along with production, from 1998 to 2008, the service area coverage of the MWA has increased from 35 to 70% of the 3,195 km^2 area of responsibility (Figure 1). The corresponding population coverage increased by about 23% from 6.4 to 7.8 million, which is almost 99% of the population in the area of responsibility. While the population coverage is based on registered population, the proportion of population coverage (i.e., 99%) is also applicable to the whole population considering that the nonregistered population are believed to be more concentrated in the urban areas where they are already covered by the MWA. The expansion of coverage has also helped to reduce groundwater withdrawal by private entities in areas that were not served earlier with piped water. The MWA has been able to persuade people to switch to tap water through reduced water charges, financial support (or subsidies) in the cost of distribution pipeline, and a reduction of the installation fee by 50%, among others (MWA, 2003).

Essential to progress in the service area and population coverage is the utility's commitment to expansion, and the implementation (in phases) of several improvement plans which are based on the original Master Plan 2017 prepared in 1990 (MWA, 1990) which is currently being revised to project future demand and supply conditions. Huge investment requirements in these plans were met with grants from the government in the earlier period (until 2000), and through the positive financial performance of the utility.

On the other hand, the MWA seems to rely heavily on the sustained availability of its water sources, though changing climate conditions may significantly affect this as the region is forecast to receive less rainfall (Bates *et al.*, 2008) and therefore reduced river flows. The revised MWA master plan, which adopts a supply-oriented approach to expand the production capacity of WTPs, indicates that a water allocation of 105 m^3/s from the

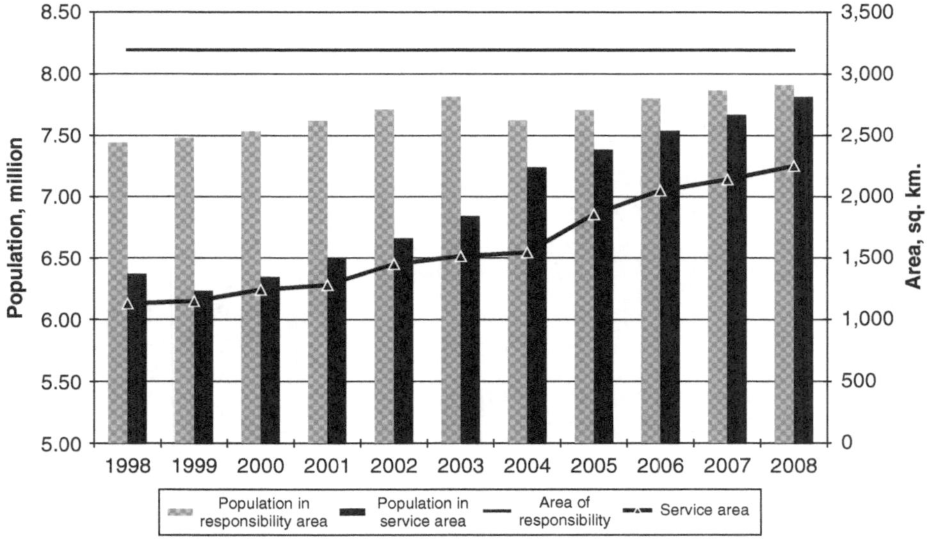

Figure 1. Service coverage of MWA. *Source:* Annual Reports 1998–2008, Metropolitan Waterworks Authority (MWA, 1999–2009); MWA (personal communication), 2009.

two rivers would be sufficient until around 2030, beyond which the MWA may need to find other sources or other ways to meet anticipated demand. Although urban and domestic water supply are considered the priority for water allocation in the country (Molle, 2001), the MWA has no direct authority on water resource planning (Suwarnarat, 2000) in the basin and the country. Thus, the supply-oriented approach adopted by the MWA should be revisited considering the limit on water allocation from its current raw water sources, and due to the challenges of the changing global climate. A balanced combination of supply and demand management should be pursued to address the impending problems of water sources and availability.

Given the extensive areal coverage (the water supply network comprises approximately 31,000 km of distribution lines with about 1.86 million connections), the MWA faces huge challenges in terms of providing a 24-hour supply of water with sufficient pressure, while at the same time minimizing unaccounted-for water. Despite the challenges, the MWA remains faithful to its commitment of providing safe and drinkable water around the clock. The results of the latest customer satisfaction survey indicate that consumers were largely satisfied with the 24-hour supply of water, which is interrupted locally only when there are breaks in the mains, other maintenance work, and during power sub-station preventive maintenance by the MEA.

In terms of water pressure in the distribution line, the MWA has been able to keep the average pressure over the entire system above its level of service of 59 kPa (equivalent to 6 m head of water). In FY2008, the average pressure throughout the service area was 6.57 m (equivalent to about 64.5 kPa) (MWA, 2009), which is higher than in the previous fiscal year. A critical factor accounting for this improved performance is the inclusion of pressure as a key performance indicator of the utility, as well as the use of 16 pumping stations spread over the network (Figure 2).

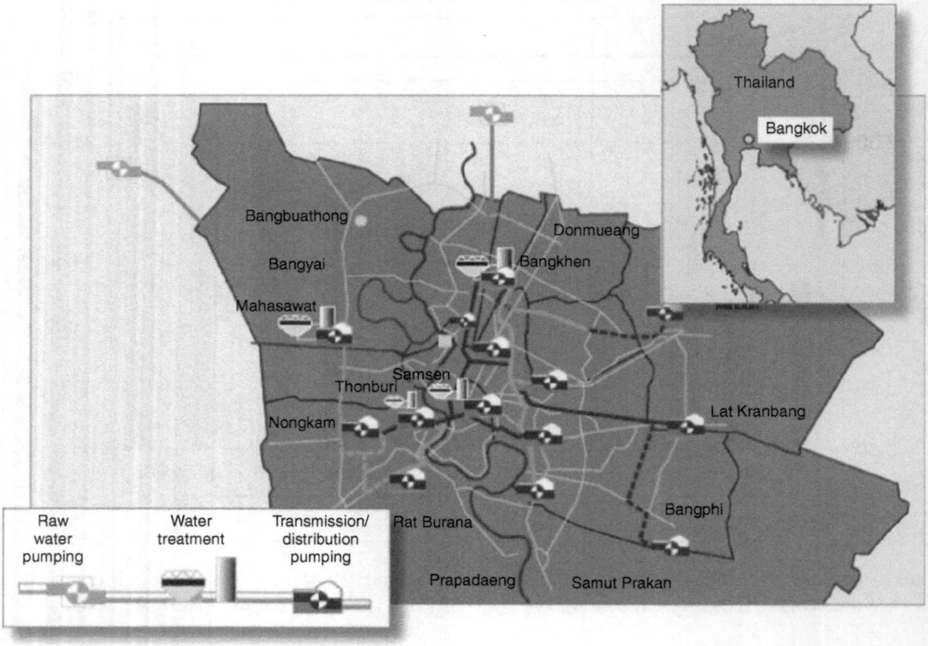

Figure 2. Major MWA water supply facilities. *Source:* ABB and MWA, 2009.

While there has been a decrease in the number of monitored locations with average monthly pressure of less than 6 m (from 11 locations in 2007 to 7 locations in 2008), this also indicates that there are still areas having an average pressure of less than 6 m, which could have encouraged the residents to install pumps and underground and overhead storage tanks. The latest customer satisfaction survey among 4,408 respondents conducted in 2008 revealed that approximately 60% are using pumps. Although it is clear that measures are required to increase water pressure, especially in low-pressure areas, the benefits of further increasing water pressure in the entire system need to be carefully investigated. This is due to the flat topography and because a significantly large number of establishments in Bangkok are buildings of more than two storeys. Further increases in pressure will benefit establishments with fewer levels as they will stop using pumps and thereby save energy costs. But how far the pressure should be increased remains a question as it could lead to increased water loss unless suitable, usually capital-intensive measures are taken, including the replacement of old distribution lines and pipes. In fact, the presence of storage tanks and pumps installed by a large number of consumers in their houses or establishments, apparently due to the height of these establishments, may merit investigation on the efficiency of providing water at relatively lower pressure to reduce unaccounted-for water. In a pilot study of an area in Bangkok, Babel *et al.* (2009) found that a 2.4 m reduction in average pressure could reduce leakage by 12.5% of the system inflow without compromising the level of service.

Leakage reduction is an important aspect of water supply management in Bangkok as there is a huge scope for improvement. In 1967 when the MWA was established, water loss was reportedly at 75% of production due to pipe leakage, illegal connections or use, and connections without meters. With reducing water loss (unaccounted for water) as one of its policies, the MWA was able to bring down water loss to 29.6% of water produced in 1991. However, water loss steadily increased again until it reached its peak of 42.14% of water produced in 1997 as a consequence of an increase in production and expansion of the service area to meet the increasing demand for water brought about by increased economic growth (Figure 3). At the same time, annual budget shortages could have led to the use of low quality pipes and affected the quality of pipe laying, replacement and repair (MWA, 1998). Based on studies conducted in 1997 (MWA, 1998) and 2000 (MWA, 2001a), the identified main cause of unaccounted for water (UFW) (81% of total loss) is leakage in the distribution system due to broken pipes or fittings. Other reasons are related to the functioning and reading of meters (14%) and illegal connections (5%).

Measures implemented by the MWA to reduce this high water loss appear to have appropriately addressed the three main causes mentioned above. In terms of leakage control, the MWA implemented the following:

- Establishment of the Water Transmission and Distribution Control Centre (using Supervisory Control and Data Acquisition, SCADA) in 1998. With the use of pressure monitoring devices installed in over 100 locations in the distribution system, as well as a remote signal system in water treatment plants and pumping stations (MWA, 2001b), the Centre collects and analyzes real time data on water flow and pressure to determine demand and subsequent adjustments needed. This resulted in faster data collection from production and distribution lines and enabled the MWA to promptly control and adjust the water distribution in accordance with water consumption behaviour (MWA, 2003). This, in effect, reduced the rate of water loss by minimizing unnecessary high flows and pressure in parts of the distribution system that do not correspond to demand.

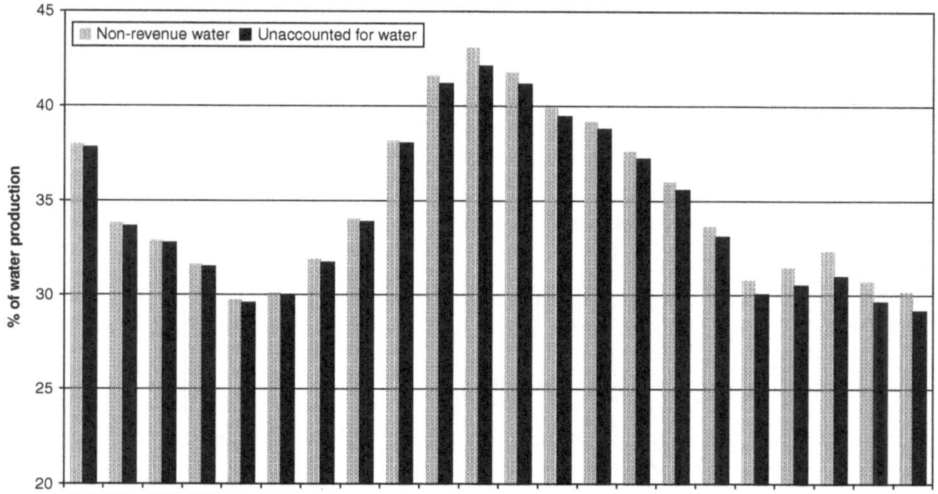

Figure 3. NRW and UFW in the MWA service area. *Source:* Annual Reports 1996–2008, Metropolitan Waterworks Authority (MWA, 1997–2009); MWA (personal communication), 2009.

- Extensive water loss reduction projects, starting in 2000. Initially implemented in the three most critical areas with the involvement of the private sector, activities included leak detection using modern technologies, and extensive pipe repairs and replacement. Based on the success of the pilot project, water loss reduction projects (distribution system improvement and leakage control) for the entire service area were subsequently implemented starting 2002.
- Establishment of District Metering Areas (DMAs) in the entire system, initiated in 2002. The MWA established DMAs covering all service areas and linked them through an efficient communication network with the control rooms at all branches in both real time and daily data input modes (MWA, 2009). This narrows down the area of responsibility for leakage monitoring and control, thereby cutting down on the inspection time necessary to solve leaking pipe problems.
- Establishment of a Water Leakage Management Application System. A system in which all information and data related to water loss management are appropriately collected and utilized in support of the policy-level and operation-level work plans to all aspects of water leakage management. This enables the MWA to carry out all operations related to leakage management (e.g., call centre, pipe repair, leakage surveys, pipeline improvements, and so on) in an integrated manner (MWA, 2009).

With respect to water meter-related factors, the MWA monitors and maintains water meters through its Water Meter Department and its branch offices. The regular replacement of meters every eight years (the average working life of meters) seems reasonable to ensure proper functioning of meters and accurate monitoring of water use. Moreover, the use of hand-held equipment for recording water consumption, along with portable printers that print invoices and bills for customers showing the water charges immediately after the reading of the water meter, provides the opportunity for customers to check their bills, and this subsequently contributes to improving the accuracy of meter readings as well as collection efficiency.

Meters are an integral part of the standard procedure for the installation of new connections and hence, connections without meters have become a thing of the past. Illegal connections seem minimal if still present in the MWA service area. This is due to the monitoring practices as well as the incentives and penalty system in place. The MWA assigns an inspection unit that conducts investigations in the field, and studies bill trends as well. It also designates a committee with the authority to implement procedures according to applicable criminal laws. Criminal cases are filed against violators, although case settlement takes time. Penalties for illegal water use are as follows:

- Misuse of water meters and water use without meters, a fine of THB 5,000–180,000 plus payment for lost water (based on estimated value of lost water at a rate of THB 20/m^3)
- Equipment damage, a fine of THB 3,000.

Individuals who report illegal connections receive 30% of the fine and the MWA officer who acts on it receives 20% of the fine as an incentive.

The above-mentioned measures contributed to a significant reduction in UFW. At the end of FY2008, UFW was at 29.17% indicating that the MWA is able to reach its target of keeping UFW at not more than 30% of produced water (MWA, 2003). This target is based on the average NRW in Asian cities (McIntosh, 2003). If estimated non-revenue generating consumption (mainly for public drinking fountains, stand posts, flushing mains, and fire fighting, which accounts for 1.4% of total consumption) is included, NRW figures for the MWA are not much different from UFW, being 30.17% in FY2008. Nevertheless, the UFW figure for the MWA is still very high and there is much scope for reduction, especially if compared with Singapore, Tokyo and Phnom Penh and other urban areas in Asia with an UFW of less than 10%. Considering that population coverage is almost 100%, the MWA should focus on reducing UFW more aggressively than its current plans of reducing it by 0.2% per year. It is suggested that MWA conducts a study to determine the economic level of leakage (ELL) and take appropriate measures accordingly.

Water Quality Management

Water quality in the distribution system has been kept within the World Health Organization (WHO) standards and the piped water has been certified safe for drinking by the Ministry of Public Health. This is due to the stringent water quality control and monitoring practices from production to distribution as well as continuous rehabilitation of the distribution system.

The MWA implements water quality control measures at three stages: raw water sources, produced water, and water in the distribution system and taps. The Agency monitors around 50 water quality parameters at these three stages with the frequency and number of samples in accordance with the WHO standards (Table 1). Analysis includes physical, chemical and microbiological parameters, heavy metals, carcinogenic compounds, and other parameters. Moreover, the MWA monitors in real-time two critical water quality parameters, residual chlorine and turbidity, at 20 locations in the system, the results of which can be viewed online at the MWA website (www.mwa.co.th).

The quality and efficiency of water treatment and monitoring measures is reflected in the international recognition the MWA has received. Two of the four WTPs of the MWA are ISO certified (9001: Bangkhen WTP in 2002, and 9002: Bangkhen WTP in 2000, Thonburi WTP in 2001) (MWA, 2002) and its laboratories received ISO/IEC 17025 certification in 2004. Moreover, results of water quality monitoring available online at the MWA website reveal that the quality of piped water is kept within international standards. This practice of on-line water quality information available to the public should help increase consumer confidence on the quality of the tap water.

However, even if the water in the distribution system is clean and safe, water from the taps may still be of inferior quality if the plumbing system in the house or establishment is not properly maintained, such as is the case in the MWA service area. Factors identified for poor tap water quality include storage tanks that are never or rarely cleaned, old and rusted pipes in the plumbing system, and filters full of bacteria, dust, and sediment (MWA, 2000). These could be the main reasons that only a small percentage of the population drink water directly from the tap—lack of confidence that it is safe enough (MWA, 1999), although surveys in recent years provide indications of other possible reasons. For example, an increasing percentage of consumers (4.2% in 2005, 4.7% in 2006, and 8.5% in 2007) suggest the need to improve water quality, particularly after pipe repairs or service interruptions. Furthermore, the near universal distrust of the quality of tap water (Tortajada, 2006a) and the concern that drinking tap water may entail significant health risks (Biswas, 2006) may have also influenced the drinking behaviour of Bangkok residents.

Table 1. Frequency of water quality monitoring

Stage	Frequency
Raw water sources	Monthly
Production/treatment plants	Every 4 hours to monthly (depending on parameter, see Dai, 1997 for more details)
Distribution pumping stations	Daily
Consumers' premises	1,200 samples per year: 100 samples are collected per month, 50 random samples from certified areas and 50 random samples from other areas
Real-time monitoring (data available online)	Sampling every 10 seconds for 2 parameters at 20 different locations within the distribution system

Source: Dai, 1997; MWA, 1998; MWA website (http://wqconline.mwa.co.th/wqc/default.aspx?uiculture=en-US); MWA (personal communication), 2009.

The MWA has responded appropriately by providing services outside its regular responsibilities to address tap water quality concerns. The Agency conducts programmes to inspect water quality in the taps of institutional establishments, both public and private, and assists them to implement corrective measures to maintain the quality of tap water. The MWA awards certificates on safe and drinkable tap water to establishments that meet the standards. From 1999 to 2008, over 1,500 establishments received such certification (Table 2).

Despite the laudable efforts of the MWA, the customer survey in 2008 showed that only a very small percentage of consumers drink water directly from the taps. Most residents (at home and/or at the office) either boil or filter tap water before drinking (71.8%); a large portion of consumers (25.8%) drank bottled water; whereas only about 2.4% drank water directly from the tap. While the percentage of those who drink tap water directly is similar to the findings of a survey made a decade ago, there is a notable decrease in the percentage of those who drank bottled water (37% in 1998), with a corresponding increase in those who filter or boil tap water (59% in 1998). This could indicate an increasing confidence in the quality of tap water.

Although only a small percentage of the population drinks tap water directly in the MWA service area, the above programmes are still noteworthy, as they would benefit the poor who may not be able to buy bottled water or filters to use in their homes. Moreover, other factors have possibly influenced the perception of consumers toward drinking tap water, and these may not be directly related to water quality. Extensive advertising and marketing by companies selling bottled water is perceived to have high influence on consumers, and the additional costs for countering this by the MWA may result in higher tariffs (MWA, 2007b). While the public generally rejects direct tap water consumption, income and education were found to have some effects on the attitudes of consumers towards other alternatives (Dai, 1997). Thus, the financial capacity of the people to purchase bottled water (which is relatively cheap compared to bottled water in developed countries) as their alternative source of drinking water cannot be discounted, considering that per capita income in Bangkok is the highest among 76 provinces, about 2.3 times the average for the entire country (Babel *et al.*, 2007). It may be interesting to determine the

Table 2. Number of establishments presented by the MWA with a Certificate of Safe and Drinkable Tap Water

Establishment	1999	2000	2001	2002	2003	2004	2005	2006	2007	2008	Total
Academic institutes	83	95	66	83	72	102	87	78	95	98	859
Government offices	0	82	29	23	24	15	16	16	16	9	231
Hospitals and medical centres	0	12	42	27	20	15	14	14	5	7	156
Villages/Religious institutions	3	0	0	1	10	24	31	28	24	22	143
Private institutions	1	4	0	0	1	1	1	5	13	25	51
Others	0	0	0	0	4	0	15	27	18	12	76
Total	87	194	137	134	131	157	164	168	171	173	1,516

Source: MWA website (http://www.mwa.co.th/download/prd01/dk_project/dk_pj_name.xls).

percentage of income spent on the purchase of bottled water in cities in developed and developing countries.

It might take a longer time for people to change their perception and behaviour towards drinking water directly from the tap given that the MWA's programme started only a decade ago. Meanwhile, measures that could help in encouraging people to drink water directly from the tap include:

- further increasing water pressure in the network so that the establishments with fewer storeys will not need storage tanks that require regular maintenance;
- continuation, possibly in a more aggressive manner, of its notable programme of assisting establishments (especially high-rise establishments) in maintaining a clean plumbing system and ensuring their participation, and
- keeping pipes clean during repairs, and corresponding efforts to prevent and control leakage and to minimize water interruptions.

Whether the production of bottled water by the MWA itself (branded as "Papa"), which is distributed to establishments such as hospitals and schools with their own labels, is affecting the MWA's drinking water campaign also needs further investigation.

Water Demand Management

Demand side management (DSM) measures being implemented by the MWA can be categorized into two: reducing UFW in distribution lines (and even in household plumbing fixtures), and encouraging water conservation among consumers. In the first component, the MWA has been able to reach its objective of keeping UFW at not more than 30%, although this figure is very high compared to other cities as earlier discussed. However, efforts on water conservation among users are apparently not successful as can be seen from domestic water consumption per capita trends, which has not changed over the last decade (Figure 4). The per capita consumption of about 210 L/day is high compared to most cities in other countries in the region and even beyond (SEAWUN, 2007; Tortajada, 2006b; McLay, 2005). The increase in the average daily total water consumption is due to population increase and increasing commercial activities such as businesses and industries. This increase in commercial activity is evident in the rise of per capita non-domestic consumption given that the served population has increased and some industries

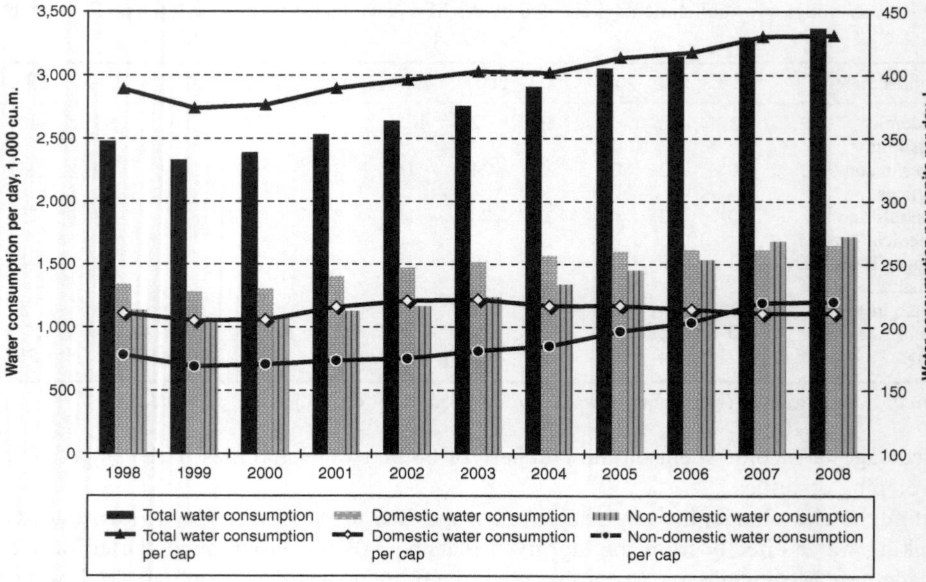

Figure 4. Water consumption by type in the MWA service area, 1998–2008. *Source:* Data collected directly from Metropolitan Waterworks Authority (MWA), Bangkok.

and other commercial establishments adopt water reuse, recycling, and saving technologies in their operations and other water use activities (Babel *et al.*, 2007, Dhungana, 2002).

While progressive rates are used in the water tariff structure to control water use (Table 3), the low water tariff does not encourage consumers to reduce water use and consume water efficiently. In addition, the tariff in the first block for residential users (0–30 m^3) does not cover the full cost of production and supply of water. For instance, the cost of water per unit sold was THB 8.69, 8.62, 8.62 and 9.16 in 2005, 2006, 2007 and 2008, respectively (MWA, 2007; 2009). However, the government wants to keep water tariffs at reasonable and affordable levels to help people with their living expenses. While the progressive tariff structure is intended to help people in the lower-income group, this also benefits those in the middle- or higher-income groups, those capable of paying for water at a higher rate but who consume water within the 30 m^3 range.

A revised mechanism for assisting lower-income people may be considered, such as the one used in Singapore whereby the government provides targeted assistance for lower-income families (Tortajada, 2006b). This is worth immediate consideration as FY2008 data show that approximately 65% of residential connections consumed less than 30 m^3 per month (accounting for about 30% of the volume of residential consumption in FY2008). Alternatively, the current first block of 0–30 m^3 per month may be divided into two or more blocks with progressive rates so that only the basic household water needs is directly supported, especially for poor people. When considering only basic household water needs, the first block may be 0–7 m^3 per month, which means about 95 lpcd for a household size of 2.5 (based on 2003–2008 population and household data from the Department of Provincial Administration for Bangkok, Nonthaburi and Samut Prakan). This figure should be reasonable for Bangkok, even with high humidity, as it is

Table 3. Current water tariff structure of the MWA

December 1999 – Present			
Residential		Commercial, government agency, state enterprise & industries	
Min. charge	THB 45.00	Min. charge	THB 90.00
Bulk	THB 10.50/m^3	Bulk	THB 13/m^3
Amount (m^3)	Rate (THB/m^3)	Amount (m^3)	Rate (THB/m^3)
0-30	8.50	0-10	9.50
31-40	10.03	11-20	10.70
41-50	10.35	21-30	10.95
51-60	10.68	31-40	13.21
61-70	11.00	41-50	13.54
71-80	11.33	51-60	13.86
81-90	12.50	61-80	14.19
91-100	12.82	81-100	14.51
101-120	13.15	101-120	14.84
121-160	13.47	121-160	15.16
161-200	13.80	161-200	15.49
>200	14.45	>200	15.81

Source: MWA, 2002.

substantially higher than the 50 lpcd that is considered sufficient (Gleick, 1996; Abrams, 2001). It is recommended that the MWA revise the water tariff such that it helps the poor while at the same time increasing revenue which can be used to drastically reduce NRW from the current level of 30% by investing, for example, in network rehabilitation and other programmes.

Recent measures implemented by the national government to assist the people with their household expenses seem not helpful to the water conservation efforts of the MWA. Due primarily to the global financial crisis experienced in 2008 and 2009, the national government implemented a subsidy programme for utilities such as electricity, transportation (bus and train), energy (cap on retail prices for cooking gas), and water. Water consumption was made free under some limit-of-usage applicable to all residential connections. It was first implemented from August 2008 to January 2009, during which connections using less than 50 m^3 are not required to pay, and if consumption exceeds the amount, they need to pay for the total amount consumed. It was extended for another period from February to December 2009, whereby connections using less than 30 m^3 are not required to pay, and lately extended to March 2010 for connections not using more than 20 m^3.

This programme has no financial impact on the MWA as the MoF reconciles the MWA accounts, and there was no observed significant increase or decrease in consumption. Nonetheless, this policy might have a negative influence on the behaviour of consumers and, consequently, on water conservation efforts. It may not be fair, however, to pin this on the MWA as the utility merely follows national government policy. It is also interesting to note the decreasing ceiling set by the government which seems to indicate that the decision is taken without a proper assessment of the required amount of water to fulfil basic human needs.

Furthermore, a DSM study commissioned by the MWA in 1995–96 recommended a number of conservation measures. Yet apart from reduction of losses, activities of the MWA seem to focus mainly on water conservation campaigns and public awareness activities.

Concrete measures to encourage water conservation, such as requiring the installation of water-efficient plumbing fixtures, is lacking. Such measures require the formulation of appropriate policies, or revision of existing laws and policies. Cooperation from relevant authorities and consumers would be important for its successful implementation. Current plumbing codes in Thailand do not specify the type of toilet and other fixtures that must be installed in new or reconstructed residential dwellings, commercial premises and public buildings (Phuaprasert, 1998). Hence, water users did not install water saving devices, given also that water fees are low and Thailand's economy was generally doing well during the past decade. This measure, however, is important, considering that domestic water consumption in Bangkok comprises about half of the total water use in Bangkok and 50% of domestic consumption is utilized in toilets and bathrooms (Otaki *et al.*, 2008).

Analysis of information further indicates that the MWA's efforts on demand management may have been constrained by the fact that the utility needs to keep its revenue high while providing water at low rates, particularly for the first block for residential connections. This is because the utility's performance is evaluated based on the profit it makes, and the bonuses and incremental salary increases received by MWA personnel, as in all other state enterprises in the country, also depends on profits—which in turn is dependent on revenue and water consumption. Thus, successful implementation of demand management measures need to be complemented with measures to increase revenue and reduce operational and maintenance (O&M) costs further so that the financial performance of the MWA would not be significantly affected and personnel continue to receive incentives that encourage increased efficiency.

Water Utility Governance

Water Pricing and Financial Resources Management

The current progressive water tariff structure of the MWA (Table 3), which has been in place since 1999, has its positive and negative attributes. It encourages consumers to conserve water as lower use is charged at a lower rate and also helps the poor people who use less than 30 m^3 per month. The MWA's commitment to help provide poor people with access to safe and affordable piped water is reflected further in the utility's effort to serve the urban poor. In particular, the MWA regulation Decree on Bulk Water Tariffs for the Poor (MWA, 2000b) sets forth guidelines on supplying piped water to the urban poor and slum dwellers. "Poor" is defined by the MWA as one having an average household income of less than THB 8,000 per month. The regulation provides details on the procedure to avail such service including the financial support and applicable rates, both for group and individual connections (Babel & Rivas, 2009). While water is provided at low rates to support the poor, the guideline also appropriately includes measures to discourage abuse or unwarranted large water consumption.

As mentioned earlier, the low rate of the first block also supports higher-income households who can afford to pay for the quantity of water consumed at a higher rate. This translates to lower than the potential revenue. Moreover, only two categories of water users are considered, residential and non-residential. The non-residential category has grouped together both commercial/industrial users and not-for-profit institutions such as government agencies, public schools and others. This clearly indicates the need to revisit and revise the water tariff structure.

With its current efforts to study and consider revisions on the water tariff, the MWA seems to be addressing the issues of high NRW, low water tariff, and water conservation. This would likely help in water conservation as consumers may subsequently reduce their consumption and probably be encouraged to install water-efficient plumbing fixtures even without the needed revision of plumbing and building codes, which could take a long time to formulate and implement.

Although the tariff structure needs improvement, and UFW is still very high at around 30% of production volume, it is interesting to note that cost recovery has been remarkable with the overall financial performance of the MWA being on a positive trend over the decade. The operating ratio (ratio of annual O&M costs to annual operating revenue) is around 0.70 (Table 4). Moreover, revenue collection efficiency, the ratio of annual collections over annual billings, is high (mostly over 100% due to arrears and penalty for non-payment of water bills on time) and account receivables for the past five years are at 0.43 month's equivalent. These figures show the financial sustainability of the utility's operations as O&M costs are comfortably covered by revenue. This is confirmed by the fact that the national government has stopped providing funds to the MWA since 2001. This indicates that the MWA has been able to compensate the weakness of its water pricing structure with other measures that brought about a positive financial performance.

Factors seen to have contributed to such a performance include financial management practices implemented by the MWA that resulted in high revenue collection efficiency and reduction of O&M costs.

The high revenue collection efficiency of the MWA could be due to the variety and convenience of water bill payment options, as well as imposition of penalties for late payments. Aside from MWA branch offices, payments for water bills may be made through the following channels:

- payment service agents or counter services (for invoices within the due date, no previous unpaid invoice, and cash payment only);
- bank account deductions (15 banks) without charges to the customers;
- at least 153 post offices within Bangkok and its periphery;
- the MWA website; and
- fixed line telephone or mobile telephone.

Consequently, customer surveys reveal that the number of respondents paying directly at MWA branch offices has decreased (from 82% in 2005 to 18% in 2008), whereas the number of those paying through counter services (such as convenience outlets and other agencies) rose from 17% in 2005 to 56% in 2008, and through bank accounts from 0.8% in 2005 to 16% in 2008. It should be noted that customers can only pay their water bills outside the MWA branch offices if they pay their bills on time, thus the survey results indicate timely payments by the customers. Penalties are imposed for failure to pay water bills on time. These include:

- water supply is cut after two unpaid bills;
- a penalty of THB 100 is charged within seven days thereafter for reconnection, after which the meter is uninstalled;
- a penalty of 25% of the new connection fee is charged for settlement within 14 days thereafter; and
- the full new connection fee is charged for cases settled beyond this time limit.

Table 4. Financial characteristics of the MWA's performance

Description	1998	1999	2000	2001	2002	2003	2004	2005	2006	2007	2008
Capital expenditure (THB million)	5,482	3,801	4,890	4,924	5,067	5,438	5,640	5,154	5,525	5,092	4,327
Annual operation and maintenance cost (THB million)	7,500	7,616	8,107	8,826	9,026	9,062	9,970	10,184	10,448	10,848	11,457
Annual operating revenue (THB million)	9,420	9,731	11,181	11,968	12,639	13,262	14,285	15,154	15,587	16,234	16,661
Operating ratio	0.80	0.78	0.73	0.74	0.71	0.68	0.70	0.67	0.67	0.67	0.69
Accounts receivable (in month's equivalent)	0.95	0.64	0.59	0.62	0.58	0.52	0.47	0.34	0.38	0.32	0.66
Annual billings (THB million)	9,427	9,919	11,426	12,041	12,706	13,350	14,382	15,157	15,659	16,448	16,912
Annual collections (THB million)	9,731	10,281	11,454	12,169	12,737	14,453	14,453	15,268	15,608	16,476	16,379
Revenue collection efficiency (%)	103.22	103.66	100.25	101.07	100.25	108.27	100.50	100.73	99.67	100.17	96.85

Note: Figures in millions have been rounded off from actual numbers.
Source: Annual Reports 1998–2008, Metropolitan Waterworks Authority (MWA, 1999–2009); MWA (personal communication), 2009.

O&M costs have decreased, apparently due to the reduction in UFW and an increase in operational and management efficiency. Increased efficiency resulted from:

- the application of a management system utilizing Enterprise Resource Planning or ERP software (the MWA started to use Systems Application and Products or SAP in 2002) and information technology;
- the reduction or cutting of unnecessary work processes;
- increased personnel efficiency; and
- adoption of outsourcing that resulted in significant decrease in staff number.

Additionally, risks associated with the currency exchange rate are reduced; the MWA policy is to depend more on local loans and revenues than on foreign loans in its investments (MWA, 2003).

Human Resources and Service Efficiency

Another important element in the positive financial performance of the MWA as well as in fulfilling customer satisfaction is the implementation of measures to increase the efficiency of staff, and subsequently higher efficiency in the delivery of services.

The MWA sets the number of staff at not more than one per 400 connections and thus it faces the subsequent task of increasing personnel efficiency to compensate for the reduction in staff number. While the number of service connections has gone up from 1,370,000 in FY1998 to 1,860,000 in FY2008, the number of MWA staff has decreased by about 25%, i.e., from 5,432 to 4,083, respectively, the latter indicating an average of 2.2 personnel per 1,000 connections (Figure 5). This is fewer than the MWA's target. This reduction in the number of staff was achieved by early retirement schemes, limiting new hiring compared to the number of retirees, use of technologies that are less labour intensive, and the adoption of outsourcing whereby some functions or activities of the MWA contracted to the private sector. Outsourced activities include gardening, security, meter reading, new service connections, system maintenance for main server computers, software monitoring and inspection for the remote area network etc.

With this reduction in staff, MWA is seen to have appropriately implemented measures to keep personnel highly skilled and efficient. These measures include:

- recruitment of new staff with high levels of education, skills, and experience;
- improvement in the technical capacities of staff through continuous training;
- implementation of a personnel evaluation system, such as the use of Performance-based Management and Competency-based Management in the appraisal of personnel performance;
- stipulation of core competency levels for every position; and
- provision of competitive salaries and benefits such as bonuses to attract and retain talent.

The expenditure for staff training in the MWA has almost tripled in the last five years indicating increased importance given to human capacity development. This subsequently resulted in the increasing number of personnel being trained, from 63% of staff in FY2003 to 83% of staff in FY2008. Moreover, the quality of training is reflected in the ISO 9001 certification received by the Human Resources Department for training and seminars.

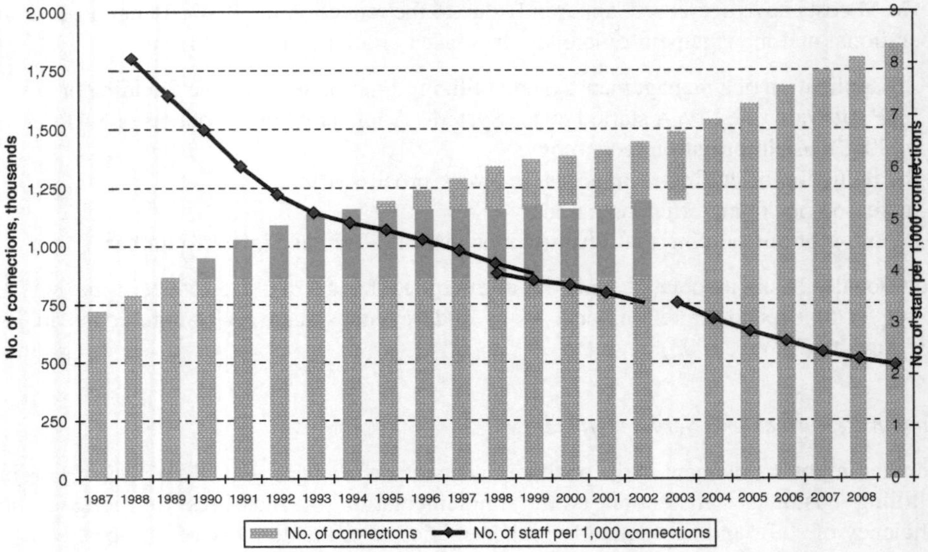

Figure 5. Human Resource Efficiency of the MWA, 1987–2008. *Source:* Annual Reports 1996–2008, Metropolitan Waterworks Authority (MWA, 1997–2009).

As for salaries, MWA personnel receive salaries much higher than their counterparts in the government sector and are competitive with those in the private sector in Thailand. For example, the average salary of the top three management personnel is currently US$63,000 per year, whereas all full-time management personnel receive salaries averaging US$35,600 in FY2008, which is at least 1.5 times the maximum compensation that can be received by government officials in managerial positions based on the Civil Service Act of 2008. Thus this salary scheme, coupled with an evaluation system for promotion of personnel and improved organizational performance, has more likely motivated MWA staff to enhance their abilities and performance.

While the number of staff with respect to the number of connections could not be taken as absolute due to the adoption of outsourcing, there is no doubt that personnel efficiency has increased given that outsourcing is normally practised to delegate other functions to private entities that have expertise on the assigned tasks or responsibilities.

As a result, the efficiency in the delivery of services has also increased. Apart from improved personnel efficiency, an essential aspect of the improvement in the delivery of services is the establishment of service standards. For instance, the MWA has set the service standard for completing new connection in three days, fixing broken pipes in 24 hours, solving metering problem in five days, low pressure in three days, turbidity/smell in three days, among others. Figures obtained from the MWA indicate that such standards appear to be satisfactorily enforced. Almost all of the complaints registered in FY2006 and FY2007 were resolved (average of 99.8% for the two years) within the duration set by MWA's standards (Table 5). Considering that the reported customer complaints in FY2003 were 183,846 (McLay, 2005), there has been a significant reduction in the number of complaints. Data for the fiscal years 2006 and 2007 reveal that most (69.3%) complaints are related to broken pipes.

Table 5. Summary of complaints received by the MWA and the MWA's response performance

Type of complaint	Received and resolved complaints, 2006				Received and resolved complaints, 2007			
	1-day	3-day	7-day	Total	1-day	3-day	7-day	Total
Broken pipe*	42,085	271	25	42,381	40,010	155	23	40,188
Metering problem**	3,815	8	0	3,823	3,551	6	0	3,557
Low pressure/ water***	12,078	24	5	12,107	15,116	38	2	15,156
Turbidity/ smell****	732	3	0	735	711	1	0	712
Unfurnished construction*****	157	1	0	158	316	0	0	316
Total	58,867	307	30	59,204	59,704	200	25	59,929

Note: Year 2008 data are not yet available due to changes in database management. Data are based on fiscal year (October of previous year to September of corresponding year). MWA Service Standard for * fixing broken pipes (except mains): 24 hrs. (1 day); ** metering problem: 5 days; *** low pressure: 3 days; **** turbidity/smell: 3 days; ***** unfurnished construction: 3 days
Source: MWA (personal communication), 2009.

Most of the complaints, along with consumer queries and emergency reports, are received through the MWA Call Centre (hotline number: 1125) based at the MWA headquarters and operating 24 hours a day. Moreover, feedback on services from consumers are also obtained through customer satisfaction surveys, which have been carried out since 1996 following the criteria for evaluation of state enterprises called State Enterprise Performance Appraisal. With the inclusion of the annual survey of customer satisfaction conducted by independent entities or consultants in the MWA Management Strategy Plan since 2008, MWA has shown its commitment to continuously improve its services to provide customer satisfaction.

It should also be noted that the division of the service area into 15 branch offices has definitely contributed to the efficiency of services as response time is significantly reduced.

Corporate Governance

The main underlying factors for the achievements of the MWA during the past decade may have been the autonomy bestowed upon the utility and its commitment to good corporate governance practices. Without these, the implementation of appropriate measures might not have been realized.

With the considerable autonomy it enjoys, the MWA has been able to implement measures to improve its performance without significant political interference. The MWA sets its water tariff as well as new connection fees independently. The MWA's current effort on considering revision of the tariff is an indication that the utility makes use of its vested autonomy appropriately.

The MWA's commitment to good governance practices is evident in the several approaches and mechanisms it has implemented. Since the mid-1980s, the MWA has adopted a business approach to its operations and management but remained committed to public service. This approach received a boost in the late 1990s until mid-2000s with the government's push for the privatization of MWA. While privatization efforts stalled in 2004, these have helped improved MWA's operations as it gears up for a more business-like approach to its operations.

An important aspect of corporate governance in the MWA is the establishment of policies and procedures to guarantee transparency in its operations and management. The main feature of this is the strong internal control system, which helps ensure that MWA personnel, units, and offices perform in accordance with regulations, operational evaluations, and financial and accounting audits, thereby contributing to financial credibility. An example is in the procurement of supplies wherein monitoring steps have been established for purchasing procedures and for treating contractors equally (MWA, 1999). With the above policies and procedure being practised by the MWA, this indicates that corruption in the utility has been minimized, if not totally eradicated. Consequently, the MWA was awarded the Best Practice on Corporate Governance Award (Transparency Management) in 2000 by the Institute of Internal Auditors of Thailand in rightful recognition of its transparent management.

Another indication of the MWA's commitment to transparency is the preparation and publication of its annual reports, which contain highlights of operations in the fiscal year, service performance, policies and work plans for the future, independently audited financial statements, and data and statistics, among others. These reports are normally issued within six months from the end of the fiscal year (October to September). Moreover, significant data and information about its services are also available at its website.

The MWA's commitment to transparency has been further strengthened with the creation of relevant committees at the Board of Directors level, namely, the Audit Committee and the Good Governance Committee. Moreover, Practice Guidelines of the MWA Audit Committee and the MWA Internal Audit Charter, which are pertinent to the standard regulations of the Committee of the Sponsoring Organization of the Treadway Commission, have been established and enforced. In 2005, the MWA received the Corporate Governance Award on Best Practice of Board of Directors from the Ministry of Finance for its good governance practices.

Another important component of governance in the MWA is its strong organizational management. The MWA continues to improve its organization in order to respond to changing needs and situations. It has adopted emerging technologies and tools to improve its organization. The MWA has improved its management system with the application of information technology and ERP software such as SAP in data processing. As such, data on the MWA's operations can be easily updated and retrieved, thereby contributing to timely and well-informed decisions.

Moreover, since 2006, the MWA has applied the principle of Economic Value Management along with quarterly Economic Profit analysis and seamless linking of financial and operational key value drivers with the Balanced Scorecard Principle (MWA, 2008). These practices helped to improve the performance of the organization. Consequently, the improvement in the MWA organization has been recognized by the national government when the MWA was awarded the Corporate Governance Award on Best Organizational Management in 2006 and 2008 by the MoF.

The authors have also noted that the MWA's achievements may not have been possible without the proper direction and guidance provided by its master plans, corporate plans and management strategies. Apart from the continuous updating of its master plan, the MWA prepares and implements medium- and long-term expansions plans, medium-term corporate plans and management strategies, and annual policies and work plans. These, we believe, provide significant contribution to the successes that MWA has achieved during the past decade.

Concluding Remarks

This study has clearly demonstrated that despite the challenges being faced by the MWA, such as a large population and area of responsibility and financial constraints, significant improvements have been observed in the provision of water supply in Bangkok and two neighbouring provinces (Nonthaburi and Samut Prakan) especially over the past decade.

While there are still areas for improvement particularly on demand management and water pricing, the MWA illustrates a good example for urban areas in developing countries facing similar challenges, as it has been able to compensate inadequacies with appropriate measures. Its notable achievements, which were made possible through its commitment to and implementation of good corporate governance practices, are evident in its financial viability and in the satisfaction of customers to its services as reflected in annual surveys conducted independently.

This paper has also shown the importance of recognizing the interrelatedness of water supply management issues. In the case of the MWA, meeting future water demand requires departure from heavy reliance on supply-oriented approaches, which in turn needs more aggressive implementation of demand management measures. The latter entails implementation of complementary measures to keep profits reasonable (e.g., further reduction of NRW, reduction of O&M costs, increase in revenue through tariff revision) so that its financial viability and the incentives received by personnel that encourage increased efficiency will not be significantly affected. In particular, the revision of the water tariff would help improve or at least maintain revenue levels with reduced per capita water consumption and hence less stress on water resources due to reduced demand and financial viability of the utility. Continued efforts to further reduce NRW would even delay investments for new water sources.

Moreover, it was also apparent from Bangkok's experience that implementation of concrete demand management measures to obtain substantial and meaningful outcomes require strong cooperation from the consumers and relevant government agencies, especially on the needed revision of existing plumbing codes.

Acknowledgement

This paper is based on a case study prepared for the Asian Development Bank in collaboration with the Institute of Water Policy (IWP), Lee Kuan Yew School of Public Policy of the National University of Singapore. The authors acknowledge the financial support provided by the ADB through the IWP to conduct the case study. We are also grateful to the MWA for providing necessary data and information for the study. The assistance provided by Ms. Nkrintra Singhrattna and Ms. Siriporn Hanmeng for coordination and data collection is also highly appreciated.

Note

1. Data from the Department of Provincial Administration (DOPA). The actual population in Bangkok, however, may not be known as many people commute to work in Bangkok or live in the city without registration.

Estimates of the nonregistered population by the Department of City Planning of BMA showed that in 2007, the nonregistered population in Bangkok was 57% of the total registered population (Panya Consultants, 2008). If this proportion is used in 2008, the total 2008 population of Bangkok is therefore approximately 8.96 million.

References

ABB & MWA (2009) *ABB drives in urban water cycle* (information brochure). Available at http://www05.abb.com/global/scot/scot216.nsf/veritydisplay/5325e072f09642a6c12571e800305fa5/$File/MWA_urban_water_cycle_low.pdf

Abrams, L. (2001) *Water for Basic Needs*, Commissioned by the WHO as input to the 1st World Water Development Report. Available at http://www.africanwater.org/Documents/BasicNeeds.PDF

AIT (1996) *Royal Initiative Projects on Water Resources Development for Metropolitan Waterworks Authority*, Final Report, Volume 1, Royal Initiative Approach on Khlong Pra Pa Improvement (Thailand: School of Civil Engineering, Asian Institute of Technology for Metropolitan Waterworks Authority).

Babel, M., Das Gupta, A., Sto. Domingo, N. D. & Kamalamma, A. (2007) *Sustainable groundwater management in Bangkok*, Case study report, prepared by Water Engineering and Management, School of Engineering and Technology, Asian Institute of Technology for the Institute for Global Environmental Strategies, Japan.

Babel, M. S., Islam, Md. S. & Das Gupta, A. (2009) Leakage management in a low pressure water distribution network of Bangkok, *Water Science & Technology—Water Supply*, 9(2), pp. 141–147.

Babel, M. S. & Rivas, A. (2009) *Case Study on Water Supply, Wastewater, and Storm Water Management in Bangkok*, Research Report prepared by Water Engineering and Management, School of Engineering and Technology, Asian Institute of Technology for the Institute of Water Policy, Singapore: National University of Singapore, Lee Kuan Yew School of Public Policy.

Bates, B. C., Kundzewicz, Z. W., Wu, S. & Palutikof, J. P. (Eds) (2008) *Climate Change and Water* Technical Paper of the Intergovernmental Panel on Climate Change, Geneva: IPCC.

BMA & UNEP (2004) *Bangkok State of Environment 2003* Bangkok: Bangkok Metropolitan Administration & UNEP Regional Office for Asia and the Pacific.

Biswas, A. (2006) Water management for major urban centres, *International Journal of Water Resources Development*, 22(2), pp. 183–197.

Cohen, B. (2006) Urbanization in developing countries: current trends, future projections, and key challenges for sustainability, *Technology in Society*, 28, pp. 63–80.

Dai, X. (1997) Evaluation of supplied water quality and consumer experience for drinking water in Bangkok Metropolis, Masters thesis, Asian Institute of Technology, 1997.

Dhungana, J. (2002) Urban water demand management in Bangkok, Thailand, Master thesis, Asian Institute of Technology, 2002.

Gleick, P. (1996) Basic water requirements for human activities: meeting basic needs, *Water International*, 21, pp. 83–92.

McIntosh, A. (2003) *Asian Water Supplies: Reaching the Urban Poor* (Manila: Asian Development Bank & International Water Association).

McLay, G. (2005) *SEAWUN Benchmarking Survey for 2003* (Hanoi: Southeast Asian Water Utilities Network).

Metropolitan Waterworks Authority (1990) *Master Plan for Water Supply and Distribution of Metropolitan Bangkok*, Final Report, Volume II (Technical Report), Bangkok: Thai DCI Co., Ltd. in association with Southeast Asia Technology Co., Ltd., Team Consulting Engineers Co., Ltd. and Safege Consulting Engineers.

Metropolitan Waterworks Authority (1997) *MWA Annual Report 1996*, Bangkok: MWA.

Metropolitan Waterworks Authority (1998) *MWA Annual Report 1997*, Bangkok: MWA.

Metropolitan Waterworks Authority (1999) *MWA Annual Report 1998*, Bangkok: MWA.

Metropolitan Waterworks Authority (2000a) *MWA Annual Report 1999*, Bangkok: MWA.

Metropolitan Waterworks Authority (2000b) *Decree on Bulk Water Tariff for Poor, No. 392/2543*, Bangkok: MWA (in Thai).

Metropolitan Waterworks Authority (2001a) *An Introduction to MWA's Unaccounted-for Water and Plan*, Bangkok: MWA.

Metropolitan Waterworks Authority (2001b) *MWA Annual Report 2000*, Bangkok: MWA.

Metropolitan Waterworks Authority (2002) *MWA Annual Report 2001*, Bangkok: MWA.

Metropolitan Waterworks Authority (2003) *MWA Annual Report 2002*, Bangkok: MWA.

Metropolitan Waterworks Authority (2004) *MWA Annual Report 2003*, Bangkok: MWA.

Metropolitan Waterworks Authority (2005) *MWA Annual Report 2004*, Bangkok: MWA.

Metropolitan Waterworks Authority (2006) *MWA Annual Report 2005*, Bangkok: MWA.

Metropolitan Waterworks Authority (2007a) *MWA Annual Report 2006*, Bangkok: MWA.

Metropolitan Waterworks Authority (2007b) Trusting the taps, as reported in *Bangkok Post*, February 4, Perspective Section. Available at http://www.mwa.co.th/ewt/mwa_web/download/prd01_e/article/bkkpost2.pdf.

Metropolitan Waterworks Authority (2008) *MWA Annual Report 2007*, Bangkok: MWA.

Metropolitan Waterworks Authority (2009) *MWA Annual Report 2008*, Bangkok: MWA.

Molle, F. (2001) *Water Pricing in Thailand: Theory and Practice*, Research Report No. 7, Bangkok: Kasetsart University, DORAS Centre.

Otaki, Y., Otaki, M., Pengchai, P., Ohta, Y. & Aramaki, T. (2008) Micro-components survey of residential indoor water consumption in Chiang Mai, *Drinking Water Engineering and Science*, 1, pp. 17–25.

Panya Consultants Co., Ltd (2008) *Climate Change Impact and Adaptation Study for Bangkok Metropolitan Region* Mid-term Report, Bangkok: Panya Consultants.

Phuaprasert, B. (1998) Water demand side management programmes for the Metropolitan Waterworks Authority, Thailand, in: *Towards Efficient Water Use in Urban Areas in Asia and the Pacific* (Bangkok: UNESCAP).

Polprasert, C. (2007) Water environment issues of Bangkok City, Thailand: options for sustainable management, *ScienceAsia*, 33(1), pp. 57–58.

Royal Thai Government (RTG) (2007) *Standard Qualification for Board of Directors and Staff of State Enterprises (v. 6) Act B.E. 2550*. Thailand: Royal Thai Government (in Thai).

SEAWUN (2007) *Data Book of Southeast Asian Water Utilities 2005* (Hanoi: Southeast Asian Water Utilities Network & Asian Development Bank).

Srivardhana, R. (1994) Water use conflicts in Bangkok Metropolitan Region, Thailand, in: J. Nickum & W. Easter, K. (Eds) *Metropolitan Water use Conflicts in Asia and the Pacific* (Boulder, CO: Westview Press).

Suwarnarat, K. (2000) Urban water problems and mitigation policies of the Bangkok megacity, *Indian Journal of Environmental Protection*, 20(8), pp. 633–637.

Tortajada, C. (2006a) Water management in Mexico City Metropolitan Area, *International Journal of Water Resources Development*, 22(2), pp. 353–376.

Tortajada, C. (2006b) Water management in Singapore, *International Journal of Water Resources Development*, 22(2), pp. 227–240.

Van der Bruggen, B., Borghgraef, K. & Vinckier, C. (2009) Causes of water supply problems in urbanised regions in developing countries, *Water Resources Management*, in press, DOI: 10.1007/s11269-009-9529-8.

[illegible]

[illegible]

[illegible]

[illegible]

[illegible]

[illegible]

[illegible]

[illegible]

[illegible]

[illegible]

[illegible]

[illegible]

[illegible]

[illegible]

[illegible]

Is Access to Water as Good as the Data Claim? Case Study of Yucatan

JORGE GUARDIOLA*, FRANCISCO GONZÁLEZ-GÓMEZ* & ÁNGEL LENDECHY GRAJALES**

*Department of Applied Economics, Faculty of Economics and Business, Universidad de Granada; **Centro de Investigaciones Regionales Dr. Hideyo Noguchi

ABSTRACT *Access to water is fundamental for people's survival. Unfortunately, statistics indicate that 13% of the world's population lack access to water. Moreover, this figure more than likely underestimates the real dimension of the problem, mainly due to definitions of access to water being too lax and not capturing the real dimension of the problem. In order to illustrate this phenomenon, an example of the situation of access to water in the Mexican State of Yucatan is examined based on original fieldwork. The study shows that the problem of access to water in this state is indeed underestimated. This is unlikely to be an isolated case, which leads us to recommend that action be taken to obtain a more accurate picture regarding the reality of access to water.*

Introduction

Water is a necessary good for the subsistence of people. The General Comment on the Right to Water, adopted by the Covenant on Economic and Cultural Rights (CESCR) in November 2002, states that "the human right to water entitles everyone to sufficient, affordable, physically accessible, safe and acceptable water for personal and domestic use" (United Nations, 2002). Of course, no individual or organization opposes a people's right to clean water. The important issue lies in how to ensure that everyone has access to clean water and proper wastewater management, taking into account the surrounding social, economic and political conditions as well as existing constraints (Langford, 2005; Al Jayyousi, 2007; Biswas, 2007).

A century and a half since the universalization of water services began in developed nations, water services are far from having spread to the world's population. The facilities for access to drinking water are by no means equal around the world. While in the Western world, access to good quality drinking water in homes is practically universal, the situation is vastly different in the majority of developing countries. According to estimates by the World Health Organization (WHO) and the United Nations Children's Fund (UNICEF), about 13% of the global population today lack access to drinking water (WHO & UNICEF, 2008). Some of the factors that result in more and more people lacking access to water on a regular and consistent basis include high population growth rates, increasing urbanization, lack of investment, inadequate levels of economic development and poor governance (Biswas, 2006; 2007).

The Millennium Development Goals (MDGs) aim to halve the proportion of people without sustainable access to safe drinking water and basic sanitation. According to official data, some progress has been made towards accomplishing this objective. However, the nature of the data may be inadequate, and therefore overestimate said progress. Official data might distort the perception of the existing problems relating to access to water. Precisely, according to the definitions that are implemented, the type of access to water and the quality of the water that reaches the end user are key aspects not shown in official statistics. This issue is even more important when we take into account its implication. The performance of statistics using definitions for access to water that lack rigor influence political decisions, as higher water access percentages in official data would lead to less political effort. By hiding the real dimension of the problem and creating a statistical illusion, the measures implemented to improve access to water sanitation are damaged.

In order to illustrate this problem, data are presented that show the situation of access to drinking water in Yucatan. Data were obtained through a field study between November and December 2008. Although official statistics show that access to water in Yucatan is almost universal at 96% (CONAGUA, 2008), the information obtained in our study reveals that the situation of access to water in Yucatan can be vastly improved. Furthermore, there are enough reasons to believe that the Yucatan case is not the only one of its kind in the world.

The article is structured as follows. The next section discusses the quality of the statistics at the global level. The following section specifically addresses the example of Yucatan. In the first part of this section, official data are used to describe access to clean water for residents of Yucatan. In the second part, the main results of the field study are shown. The final section summarizes and concludes.

Questioning Statistics on Water Access

The United Nations Water Conference, held in Mar del Plata in 1977, declared the period 1981–90 as the International Water Supply and Sanitation Decade (henceforth, the Decade). The objective of the Decade was to provide clean water and sanitation to every person in the world by 1990. Although it did not reach its goal of universal access by 1990, it did have a great impact as it put water and sanitation issues in the global agenda. In fact, hundreds of millions of people all over the developing world received access to clean water, which would not have occurred without this proclamation (Biswas, 2006; 2007). Some years later, the MDGs, more specifically Target 7c, contemplate the objective of halving the proportion of people without sustainable access to safe drinking water and basic sanitation by 2015, based on the situation in 1990. The MDGs have therefore partially taken up the challenge of the Decade, but this time with the less ambitious objective of halving the percentage of people affected instead of the making access to clean water universal.

According to the latest data, the world is on track to reach the target for drinking water. The Joint Monitoring Programme (JMP) report that relies on the data provided by governments states that 87% of the global population have access to improved drinking water sources, in comparison to 77% in 1990 (WHO & UNICEF, 2008). In addition to this, current trends suggest that more than 90% of the global population will use improved drinking water sources by 2015 (UNESCO, 2009). These results may seem to be good news *per se*, as they imply that access is being improved, and in a not very optimistic scenario this could lead to hope that the objective of universalization could cease to be a utopia in the future. However, we must examine the data to ascertain whether these results are in fact as positive as they seem.

Taking that into consideration, the quality and trustworthiness of the data is under question. An important and prominent issue underlying this involves the definitions that said data are based on; that is, what the definition of access to improving drinking water is. The JMP recognizes that the definitions of access can vary widely within and among countries. As a result, this programme, which aims to analyse and report data and trends at the global level and across time, creates a series of categories for 'improved' and 'unimproved' facilities. It considers that an improved drinking-water source is "one that, by nature of its construction or through active intervention, is protected from outside contamination, in particular from contamination with faecal matter".

This definition is problematic for several reasons. Firstly, it allows us to question how such a significant increase in access to improved drinking water is possible without a similar advance in sanitation systems. If no such advance is made, the progress towards improving drinking water could be considered unsustainable over time. According to the most recent data, sanitation improvements have not been as marked as expected within the framework of the MDGs. Between 1990 and 2006, the proportion of people without improved sanitation decreased by only 8%, reaching a level of 62% (WHO & UNICEF, 2008). Without a great acceleration in progress, the world will not achieve the MDGs' sanitation target by 2015. This means the good news regarding access to improved drinking water is not as good as it initially seemed, as progress in both access to water and sanitation must be made.[1]

Secondly, the MDGs also do not guarantee that water is safe for drinking. In contrast, the definition used by the objective in the Decade states that "everyone should have access to clean water that is drinkable without any additional treatment". These definitions are very different and the MDG definition results in progress towards the objectives of access to water being overvalued. The discussion then moves mainly to the provision of a certain quantum of water, irrespective of quality in terms of drinking. Furthermore, the physical presence of an improved water source is not always an accurate indicator of increased access, because technologies do not always work correctly. One example of this are water cuts. Many people run out of water for long periods of time and are forced to go to irregular water markets or other unsafe alternatives (UNDP, 2006). Millions of poor households, therefore, regularly use both improved and non-improved water sources, even when they are recognized as households that have access to improved water. The fact that they have to look for alternatives such as rivers and streams, and that the wastewater system does not work properly, further aggravates the problem of access to water that it is not captured by statistics.

These two reasons should not be taken lightly, because they result in the conception of clean water and proper wastewater management being far from their consideration as a human right. Article 11 of the General Comment of the CESCR states:

> The adequacy to water should not be treated narrowly, by mere reference to volumetric quantities and technologies. Water should be treated as a social and a

cultural good and not primarily as an economic good. The manner of the realization of the right to water must also be sustainable, ensuring that the right can be realized for present and future generations. (United Nations, 2002)

Another reason that explains the lack of capacity of existing data to explain the reality of the water scenario is the methodology used in surveys, more specifically their coverage. In national surveys, many poor people are not included, as they live in areas that are not officially recognized by governments. These areas normally suffer from highly deficient access to water and sanitation.

The fact that progress in access to water is overvalued raises serious doubts as to the quality of the statistics employed. A phenomenon of statistical illusion could be preventing us from seeing that a higher percentage of the world population lacks access to water under conditions to which every person is entitled. This lack of transparency is perverse, as statistics are used as a basis to design internal policies in each country, as well as international policy regarding the evolution of the problem. It is beyond any doubt that to consider households as having access to water as defined in the framework of the MDGs does not guarantee that their members will avoid the diseases associated with low quality. Nor does it contribute to gender equality in the developing world, as having water in the household would erase the need to search for water elsewhere, a task that is normally carried out by women and young girls. Moreover, the definition does not imply that supply is constant throughout the day. As a result, governments, non-government organizations and international organizations lack a solid basis for action. There is not only a statistical illusion in the case of access to clean drinking water for consumption, but also in sanitation. Studies carried out in Mexico by the Third World Centre for Water Management for the Inter-American Development Bank demonstrated that if access to proper wastewater management is considered, only about 11% of the population in Latin America was covered in 1990 (Asian Development Bank, 2007; Biswas & Seetharam, 2008).

The trend in access to water and sanitation for several countries is displayed in Table 1. These figures contrast with the conditions in several countries where the situation is worse than the data would have us believe. Reality is therefore more complex and more dynamic than suggested and reflected by the situation of several cities in some countries. The United Nations Development Programme (UNDP) 2006 report provides some examples that are worth bearing in mind (UNDP, 2006). In Dar es Salaam, Tanzania or Ouagadougou, Burkina Faso, around 30% of households are connected to a public water supply. In Kenya, access to water rose from 41% to 57% and sanitation from 39% to 42%, but population density, crowding and a lack of infrastructure have created a desperate situation in cities such as Kubera. In Lahore and Karachi, cities in Pakistan where around half the population lives in poor and illegal settlements, epidemics transmitted by water are commonplace. In those cities, 40% of the water supplied is not filtrated and 60% of residuals are not treated at all. However, access to water in Pakistan stands at 90% and access to sanitation at 58%, according to official data.

In Mexico, access to water increased from 56% to 81%, and in the state of Yucatan 96% of the population are estimated to have access to water, defined as the percentage of households connected to the water service. As an additional example of statistical illusion, the next section will elucidate if there is also disparity between official data and reality in the case of Yucatan.

Table 1. Access to water and sanitation according to official statistics in several developing countries

	Improved water source (% of population with access)		Improved sanitation facilities (% of population with access)	
Country Name	1990	2006	1990	2006
Algeria	94	85	88	94
Argentina	94	96	81	91
Bangladesh	78	80	26	36
Bolivia	72	86	33	43
Botswana	93	96	38	47
Brazil	83	91	71	77
Cambodia	19	65	8	28
Cameroon	49	70	39	51
Chile	91	95	84	94
Colombia	89	93	68	78
Cote d'Ivoire	67	81	20	24
Ecuador	73	95	71	84
El Salvador	69	84	73	86
Ethiopia	13	42	4	11
Ghana	56	80	6	10
Guatemala	79	96	70	84
Haiti	52	58	29	19
India	71	89	14	28
Kenya	41	57	39	42
Liberia	57	64	40	32
Madagascar	39	47	8	12
Mexico	88	95	56	81
Mozambique	36	42	20	31
Nicaragua	70	79	42	48
Pakistan	86	90	33	58
Paraguay	52	77	60	70
Peru	75	84	55	72
Rwanda	65	65	29	23
Senegal	67	77	26	28
South Africa	81	93	55	59
Sudan	64	70	33	35
Tanzania	49	55	35	33
Turkey	85	97	85	88
Uganda	43	64	29	33
Zimbabwe	78	81	44	46
Burkina Faso	34	72	5	13

Source: WHO/UNICEF Joint Monitoring Programme. Data extracted from World Bank World Development Indicators data on the Millennium Development Goals.

The Case of Yucatan, Mexico

Water in Yucatan: A General Description

This section gives an overview of the context in which the study was carried out. The specific issues that are dealt with include the geographic situation and natural conditions of the region, agencies responsible for water management at national and regional level, and the water balance in the area.

Geography, climate and hydrology. Yucatan is one of 31 states that, along with the Federal District, make up the 32 Federal Entities of Mexico. It is located in the southeast of the country on the peninsula of the same name (Figure 1) and is divided into 106 municipalities. The population is approximately 1.7 million. Merida and the metropolitan area alone account for more than half the population of the entire state. It is estimated that 83% of the population live in urban enclaves and the other 17% live in rural areas.

The state of Yucatan is one of the most prosperous areas of the country, largely due to the vigorous tourism industry. The archaeological site of Chichen Itza, vestige of the Mayan culture and named one of the Seven Wonders of the Modern World, is the foremost tourist destination. However, there are still great disparities between different population classes. While the coast is home to the most prosperous areas, according to the National Council for the Political Evaluation of Social Development (CONEVAL), inland areas in the southeast region contain municipalities with a high poverty rate (CONEVAL, 2007). These areas have problems relating to hunger, education, health, basic services and housing quality and space.

The climate is predominantly hot and humid. The average temperature is about 26°C year round. Yearly average rainfall is between 1,000 and 1,400 mm, with a tropical rainfall

Figure 1. Location of the state of Yucatan. *Source:* National Institute of Geography & Statistics (INEGI).

period from May to October. The most concentrated rainfall takes place in June and July and there is a relatively dry spell between April and May.

Yucatan does not have surface currents of water. The hydrology of the state is primarily groundwater. Most water resources come from cenotes, estuaries, lagoons and ponds. The high rainfall during certain months of the year, the great infiltration capacity of the soil and the low topographic gradient favour subterranean water replenishment.

Administrative management of water at the regional and national level. The National Waters Act establishes that the exploitation or excessive use of water should be regulated through concessions assigned by the Federal Executive branch through CONAGUA. The licenses are registered at the Public Registry of Water Rights (REDPA).

CONAGUA is the public entity responsible for administrating and preserving water throughout Mexico, as well as maintaining the sustainable use of water resources. This agency belongs to the Ministry of Environment and Natural Resources. To comply with the laws set forth in the National Waters Law, CONAGUA is divided into three large areas: Central Offices, Basin Organizations and Local Management.

The Basin Organizations are responsible for administering and preserving the waters in each of the 13 administrative regions into which the country is divided. The Basin Organization of region 12 has offices all over the Yucatan Peninsula, covering the states of Campeche, Quintana Roo and Yucatan. Among its various duties, it determines the availability of water, achieves an efficient use of water resources, assures the preservation of aquifers, guarantees the quality of surface water, and promotes the correct use and preservation of water. One peculiarity from a managerial perspective is that the Basin Organizations in the Yucatan Peninsula have no authority over large dams. This is because almost all the water collected comes from under the ground.

Uses of water. The water resources used by the State of Yucatan come from the Yucatan Peninsula Aquifer. Considering the net average refill rate of the aquifer, its capacity is 7271.2 million m^3 per year. For the three states on the Peninsula, the volume of water licensed and registered at the REDPA is 2,181.5 million m^3 per year (CONAGUA, 2008). Unlike the situation in some aquifers in the Centre and North of the country, the Yucatan Peninsula aquifer is not overexploited. In fact, the level of water stress in the administrative region is one of the lowest in the country.[2]

The volume of water granted under licenses in the state of Yucatan is over 50% of the total allowed by the Basin Agency in the entire Peninsula. The volume of water devoted to agriculture in the state is over 74%. This volume would cover an irrigation district[3] and 1024 irrigation units for a total area of 45,421 hectares of irrigation. A deficient canal system, which has been deteriorating for a long period of time due to a lack of economic resources, as well as widespread gravity irrigation, has led to an inefficient use of water on behalf of agriculture. Public supply and domestic use account for 22% of water consumption.

Water Coverage in Yucatan

Pursuant to Paragraph III of Article 115 of the Political Constitution of the United Mexican States, municipalities have jurisdiction over public services including drinking water, drainage, sewage and the disposal of rainwater. However, as is natural, the management of these services is delegated to other agencies.

The Yucatan Water and Sewage Board (JAPAY) is the public agency responsible for the administration, operation, maintenance, expansion and construction of drinking water and wastewater systems throughout the state of Yucatan. JAPAY objectives are to guarantee a quality resource and efficient water distribution to all areas in the state.

According to official data, the service coverage of the water supply is 96%. In urban areas it is almost 97%, while in rural areas it stands slightly below 94% (CONAGUA, 2008). All of the water comes from below ground and is therefore less exposed to surface water pollutants, which might be a reason for the state not having a water treatment plant.

Figure 2 shows changes in the state of Yucatan from 1990 to 2005, showing total residential housing with total water coverage according to official figures. It is noted that in those 15 years, there has been a substantial effort to expand service coverage. It should be borne in mind that an upward trend in the number of inhabited houses increased the service area during the period. The number of households connected to the water service in relation to total households has improved substantially in the last few years, rising from 71.5% to 96%.

Official statistics show that water service has become practically universal. However, this information should be interpreted with caution. The criteria given to present the data do not inform about the type of access or the quality of clean water, issues the present authors wish to emphasize.

Firstly, clarification is needed of what CONAGUA/INEGI understands as access to water when processing and gathering data to build statistics. According to CONAGUA, "Water coverage includes people who have water piped either inside their house; outside their house, but inside their property; to a public tap; or through other housing" (CONAGUA, 2008, p.78). In Yucatan, each household has a plot of land upon which their house is built, although very often the house is not built right on the edge of the street, but is set back a few meters. When drinking water pipes are laid along the street, this is outside the home but inside the lot. If a household wants access to water from inside the house, a plumber must be paid to carry out the necessary installation. Figure 3 provides a spatial

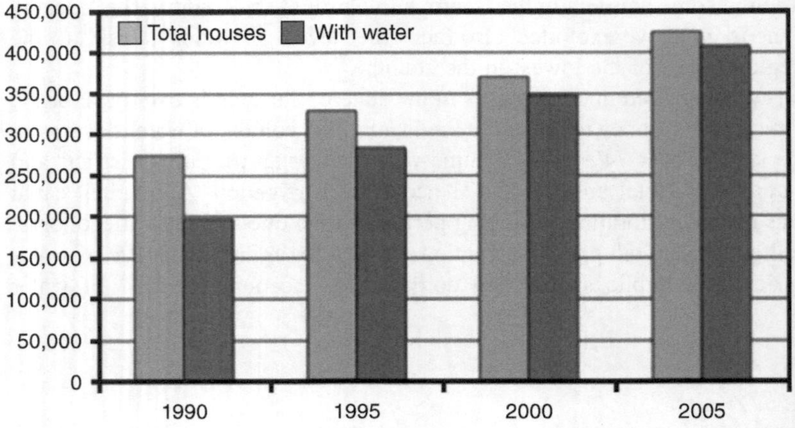

Figure 2. Water coverage in Yucatan, 1990–2005. *Source:* XI and XII Population & Housing General Census, 1990 & 2000 database; I & II Count of Population and Housing, 1995 & 2005 database.

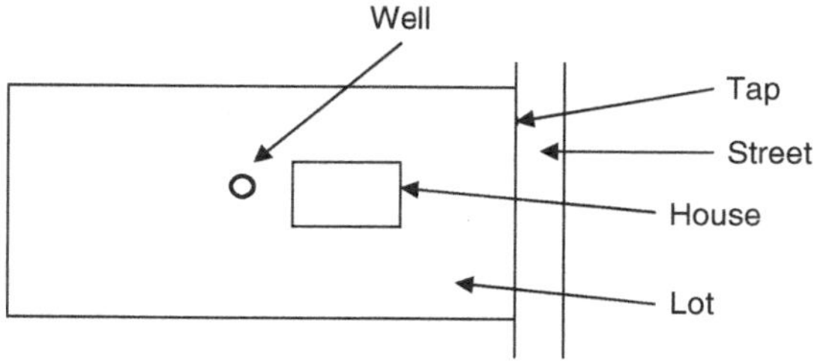

Figure 3. Spatial distribution in a Yucatecan home. *Source:* Authors' rendering.

distribution of a Yucatecan home. Sometimes houses do not have access to the water supply pipe. These are normally houses that are far from the distribution tank in the community. Therefore, they must rely on a public tap or an external house. Obviously, availability is clearly not the same for those who have to go out of their houses to access it. It would be more appropriate to distinguish between different types of access to measure coverage.

Secondly, and more importantly, although the population has water service, CONAGUA acknowledges that coverage does not necessarily mean that drinking water is available. The water that inhabitants have access to may not be of sufficient quality to be considered potable (CONAGUA, 2008, p.78). Not all of the water supplied meets the necessary standards to ensure that it is fit for consumption. Moreover, although groundwater is less exposed to pollutants, significant risks remain.[4] It is therefore obvious that a lax definition of access to water in Mexico has led to the underestimation of the problem. If the definition used by the Decade was employed instead, the data would reflect the situation more accurately and access to water coverage would be lower than current official statistics indicate.

In fact, some common solutions in Mexico are to consume bottled water, or to boil, chlorinate or filter water using membrane technology. Indeed, according to INEGI, sales of purified water increased by almost 2 million gallons between 2001 and 2008. This means that the sales volume of purified water during this time increased by 53.8% (INEGI, 2009). In 2008 Mexico was the second largest consumer of bottled water in the world after the United States, and topped the ranking of consumed bottled water per capita, with 224 litres per person (Beverage Marketing Corporation, 2009).

Water treatment within the household, the purchase of bottled water, and the analysis undertaken on the quality of the water are evident signs that clean water is not reaching households. Moreover, a lack of water quality causes such illnesses as infectious intestinal diseases in Yucatan. There are problems of child morbidity resulting from diarrhoea caused by unsafe drinking water.[5] In the following section we will attempt to complete the information from the official statistics with the information obtained from our field work.

Water Access in Yucatan According to Fieldwork

This section presents an analysis of the primary forms of access to water obtained from field work conducted in the State of Yucatan from November to December 2008 with information from 373 households.[6] The information obtained, regardless of the aggregate data from CONAGUA, provides a more accurate idea of the situation regarding access to water in which the population of Yucatan finds itself, and ultimately reveals some of the key factors to take into account in order to improve service.

Types of access. The research gives information on the type of access in the state of Yucatan, making a distinction between access to water in the house and access to water outside the house, either from a tap or a well. Information is presented in Figure 4.

The first relevant fact is that, taking into account a similar definition to that used by CONAGUA, the access to water figure supplied by this institution is very similar to that obtained in the research.[7] A high percentage of households have access to water, up to 97%. However, out of all the households with access to water, only 48% had access within the home. In 52% of cases, access to water is from outside the home. The main alternative to not having access to water inside the home is access to water through a public tap.

On the other hand, households that have wells typically also have access to water through taps, either inside or outside the home. In the sample, only nine households had access to water exclusively through a well. In most cases wells are used as a supplement to the tap, never as a substitute. Instead, their use is reserved for cases of supply problems in the network. Nearly half of the households interviewed said that cuts in supply were relatively frequent. As for the consistency of the supply, just over one in five households complained that cuts were frequent. The percentage of the total sample of respondents who claimed to have suffered cuts very often, often, or sometimes amounted to 53.8%.

Considering the different comfort levels for each type of access, one could assume that the network within the home is preferable to other types of access, and having the network outside the home is preferable to a well. Relating the type of water access to the level of household income shows that access to water is a sign of inequality between families. Measuring the level of income by quartiles of the variable, we see that there is a direct relationship between this and the degree of access to water in the home. Water inside the

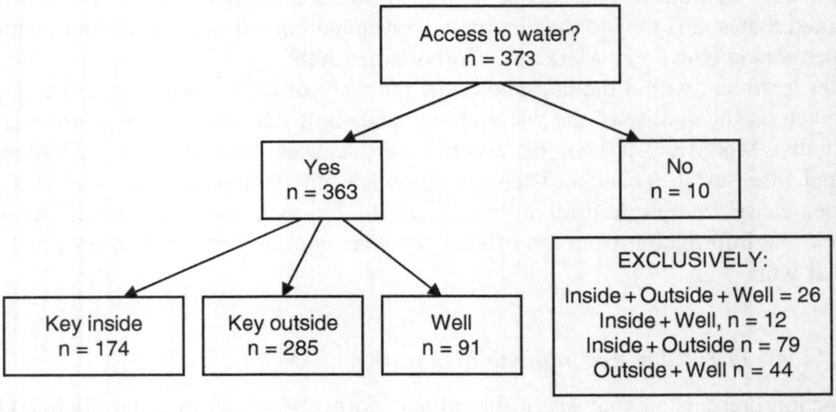

Figure 4. Characteristics of access to water by households. *Source:* Drawn from the field study.

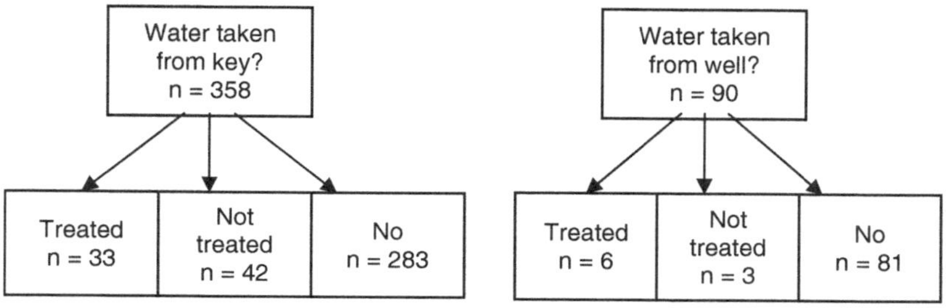

Figure 5. Water treatment according to the source of the same. *Source:* Drawn from the field study.

house is available in 34.1% of households belonging to the lowest quartile, while in the highest quartile this figure rises to 65.3%.

Intake and Water Perception

In official water coverage statistics, CONAGUA recognized that not all households have a supply of drinking water. Consequently there is an undetermined percentage of households with access to water that is not fit for consumption. A certain amount of water that reaches the houses has not received the necessary treatment to be considered drinkable.

Information obtained through this field work is an approximation of the scale of the problem in Yucatan. The interviewer asked if water from the network was consumed, and if consumed, whether the water was subjected to any treatment prior to consumption. Figure 5 shows the households surveyed where the residents treat water for drinking, those that drank the untreated water, and those which do not drink the supplied water.

The first striking fact is that most households do not drink the water supplied by the company managing the service. About 80% of households which have access to water through a tap, do not consume that water. The percentage of households which do not consume that water rises to 90% when it is derived from wells. These data are consistent with the fact that 88.5% of households surveyed bought bottled water. As an additional indication of poor water quality, we should say that of the few homes that drink water supplied by the operator of the service, many treated their water. The low percentage of households in which water is consumed, as well as the number of households that treat water before drinking, are indications that said water is unfit for human consumption. We see clearly at this point that if the definition of the Decade, "everyone should have access to clean water that is drinkable without any additional treatment", were used, the figures suggesting almost universal access in Yucatan would be a vast exaggeration.

Data from the fieldwork include information about the perception of water in homes. The information obtained has been developed into a simple subjective index of water perception. The index is a weighted assessment given by the respondents in terms of the colour, smell and taste of the water.[8] The greater the index, the higher the perception of quality. This index has an average of 6.81 out of a maximum score of 12.

Table 2 compares average perception according to key variables, and then uses a t-test to estimate if these measures are different under the assumption that the variances are unequal. According to the results of Table 1, we can conclude the following:

1. Households which consume untreated water (11.5% of the sample) have a higher perception of quality.

Table 2. Mean difference between the perception of water considering certain key variables

	Yes	No
Drink without treatment?	7.42	6.73
Have a well?	6.41	6.93
Buy purified water?	6.74	7.40

Note: The averages of the key variables are different with a confidence level of 95%. Inequality of variance was assumed and that variable perception is distributed normally (test of normality indicates so at a confidence level of 95%).

2. Also as expected, it was observed that the households which buy bottled water (88.5% of the sample) have a lower average perception than households that do not buy water. Average spending on bottled water for all households amounted to 150 pesos per month (about 11.5 dollars at the official exchange rate when the data was taken) with the average percentage of income spent on bottled water being 4.5%. The perception of water is lower in households with access to water from wells.

Given the evidence on non-drinking water, the question raised is why some households drink water that has not been treated. One possible explanation may be a lack of resources to buy bottled water or purchase products to treat water. To check the influence of deprivation, we compared the lack of water treatment to those households that reported that they had failed to satisfy their nutritional needs at least once in the last month. The result of a chi-square test confirms the existence of a significant relationship between both variables. The 58.1% of the sample who drink untreated water have gone hungry, compared to a rate of 35.6% that have gone hungry and drank purified or treated water.

Another possible explanation for this phenomenon is the lack of education and ignorance that this water can be harmful to human health. Data analysis using a chi-square test confirms the existence of a significant relationship between people who cannot read and write and those who use untreated water. Of all the illiterate people who answered the survey (31), 22.6% did not treat water, compared to 10.5% who said that they can read, and that they treat water or do not drink it.

Conclusions

Water is necessary for life. The UN recognizes that access to sufficient, affordable, physically accessible, safe and acceptable water for personal and domestic uses is a fundamental right of every person. However, we are far from being in this situation today, as there are large global imbalances in relation to access to water. While in industrial countries, access to water is nearly universal, in the rest of the world there are areas where deprivation is severe. In spite of water access data showing an improving situation with respect to the achievement of the MDGs, a phenomenon of statistical illusion seems to be resulting in an underestimation of the problem. According to data available from the JMP for Water Supply and Sanitation (WHO & UNICEF, 2008) many countries have clearly improved their access to water in recent years. Although the data are correct under official definitions, the information provided may be insufficient to ascertain the real situation of access to water in each country. Lax definitions in access to water data gathering result in published statistics considering households as having access to water when they really lack "access to clean water that is drinkable without any additional treatment". The reality is more complex and therefore it would be desirable to have more detailed data that capture the problem more accurately.

To illustrate this fact, information from a recent study conducted in the State of Yucatan was examined. Data from the field study reveal characteristics of these accesses to water which are not reflected in official statistics. Although according to official data, water service is virtually universal, data from fieldwork make it clear that more than half of all households get their water from outside the home, more than half of the households surveyed said that the supply is often blocked or cut, and more strikingly, of all the households with access to piped water, only 20.9% drink the water supplied by the company managing the service. Most of the population with access to water (88.5%) buy bottled water for consumption. This indicates that the perception of the quality of the water is low to the point that it is not fit for human consumption, and as such, the quality of the service should improve.

Official statistics maintain that 97% of the population has access to water in Yucatan. However, in view of the field data obtained, the following questions should be seriously considered: Can we say that in Yucatan, despite widespread availability and low water stress in the state, there is a universal water service that meets the standards maintained by the United Nations? Is water physically accessible, safe and of acceptable quality for personal and domestic use in Yucatan? While this appears to be the case when looking at official data, it is clear that much remains to be done to improve the water service. Unfortunately, Yucatan does not seem to be an isolated case. There appears to be a phenomenon of statistical illusion, stemming mainly from the use of lax definitions of access to water, which tends to underestimate the scale of the problem. The authors recommend using definitions of access to water that can better reflect the reality and the dimension of the problem. Therefore, when talking about access to water in a household, we are referring to access that guarantees the health and the well-being of household members, which means political action can be implemented using a precise reality as a basis.

Acknowledgements

The authors would like to thank the helpful comments made by the assistants in the 1st Workshop on the Economics of Water, held in Granada the 4th of September 2009, especially the comments made by Asit K. Biswas and Cecilia Tortajada. Ángel Lendechy Grajales acknowledges the internal support from the Centre of Regional Research "Hideyo Noguchi", Universidad Autónoma de Yucatán. Francisco González-Gómez and Jorge Guardiola acknowledges the economic aid received from the Spanish Ministerio de Ciencia e Innovación (project ECO2009-08824, subprogram ECON). The usual disclaimer applies.

Notes

1. The reason for this is that, especially in urban areas, the improvement of the water supply introduces new water, and the majority of this becomes wastewater afterward. This wastewater must be collected, treated and disposed of in a way that does not harm the environment. Provision of the urban water supply, even if it becomes universal, will not be sustainable by itself unless adequate arrangements can be made for wastewater collection, treatment and disposal (Biswas, 2006).
2. The water stress level is calculated by the quotient between total water volume concessions and the median availability of water. Water is considered to be under strong pressure when the ratio exceeds 40%. In the case of the Yucatan Peninsula, this figure stands at 7.2% (CONAGUA, 2008).

3. Irrigation districts are irrigation projects designed by the federal government including works for the storage and distribution of water.
4. The major risk of groundwater contamination in the state of Yucatan stems from several factors, including the natural characteristics of the ground, the lack of sewage and septic tanks, cesspools and drains, which facilitate filtration almost directly into the water table. Moreover, the fissured nature of the subsurface accelerates the transfer of water into the water table as the unsaturated or vadose zone has almost no attenuation capacity, particularly in regard to microbiological contamination, as the openings are larger than the pathogenic microorganisms (Pacheco, Cabrera, Perez, 2004). Another characteristic of coastal pollutants in the state is salt water intrusion (Paredes, 1997). A study by Pacheco Cabrera and Pérez (2004) in which samples of water in several communities were taken, found that the quality of the groundwater in Yucatan could be classified as bacteriologically "dangerous" and "heavily contaminated". In the eastern part of the state, this situation constitutes an enormous risk of gastrointestinal disease through direct consumption of water.
5. According to the Yucatan Department of Health, infectious agents that cause diarrhoea are usually transmitted via a faecal-oral route. The causative agents are viruses, bacteria and parasites and the most common transmission mechanism is through the consumption of contaminated food or water. According to the Department of Health, intestinal infectious diseases decreased by 5.4 percentile points between 2006 and 2007. Several government programmes have influenced the reduction of mortality rates associated with infectious intestinal diseases in children under five years of age from 122.7 deaths in 1990 to 14.7 in 2006 (INEGI, 2009). However, there is still a great deal of room for improvement.
6. To determine the sample, the procedure was as follows: first a municipal sample was obtained, then a sample of towns and finally homes were selected using data from INEGI. In the first stage, 51 municipalities were selected at random. In the second stage, 39 towns were selected at random from the total within the range of 500 to 15,000 houses. Finally, 373 houses were selected from these towns using their maps. The sample is representative of the selected universe. The municipalities were located across the state of Yucatan. Sample sizes were calculated using the Epiinfo 6.01 program.
7. The definition of the fieldwork did not take into account those households that had access to water through a public network or through an external house.
8. Survey respondents rated the perception of water considering various aspects (colour, smell, taste) on a scale of 1 to 5, with 1 being very good and 5 very poor. The remaining values are intermediate in perception, with 3 considered as normal. The index is constructed as the sum of the three variables, creating a discrete quantitative variable with a range of 3 to 15, both inclusive. The perception indicator is calculated by rescaling the above and subtracting 3 units to start from zero, then calculating 12 minus the previous variable, so that 0 would be the lowest level of perception and 12 would be highest in the perception scale.

References

Al Jayyousi, O. (2007) Water as a human right: towards civil society globalization, *International Journal of Water Resources Development*, 23(2), pp. 329–339.

Asian Development Bank (2007) *Asian Water Development Outlook, 2007—Achieving Water Security for Asia* (Manila: Asian Development Bank).

Beverage Marketing Corporation (2009) *The Global Bottled Water Market*. Report (New York: Beverage Marketing Corporation).

Biswas, A. K. (2006) Water management for major urban centres, *International Journal of Water Resources Development*, 22(2), pp. 183–197.

Biswas, A. K. (2007) Water as a human right in the MENA region: challenges and opportunities, *International Journal of Water Resources Development*, 23(2), pp. 209–225.

Biswas, A. K. & Seetharam, K. E. (2008) Asian water development outlook, 2007: achieving water security for Asia, *International Journal of Water Resources Development*, 24(1), pp. 145–176.

CONAGUA (2008) *Statistics on Water in Mexico* (Mexico DF: Ministry of Environment and Natural Resources).

CONEVAL (2007) *The Maps of Poverty in Mexico* (Mexico DF: CONEVAL).

INEGI (2009) *Statistics about World Water Day: National Data* (Mexico DF: INEGI).

Langford, M. (2005) The United Nations concept of water as a human right: a new paradigm for old problems?, *International Journal of Water Resources Development*, 21(2), pp. 273–282.

Pacheco Ávila, J., Cabrera Sansores, A. & Pérez Ceballos, R. (2004) Diagnosis of groundwater quality in the municipal supply systems in the state of Yucatan, Mexico. *Engineering*, 8(2), pp. 165–179.

Paredes, A. J. (1997) Water management in Mexico: a framework, *Water International*, 22, pp. 135–139.

UNDP (2006) *Beyond Scarcity: Power, Poverty and the Global Water Crisis.* Human Development Report 2006 (New York, NY: UNDP).

UNESCO (2009) *Water in a Changing World.* The United Nations World Water Development Report 3 (Paris: UNESCO).

United Nations (2002) *Substantive Issues Arising in the Implementation of the International Covenant on Economic, Social and Cultural Rights*, General Comment No. 15 (New York, NY: United Nations). Available at http://www.unhchr.ch/tbs/doc.nsf/0/a5458d1d1bbd713fc1256cc400389e94?Opendocument

WHO & UNICEF (2008) *Progress on Drinking Water and Sanitation: Special Focus on Sanitation* (Geneva: World Health Organization).

[illegible]

[illegible]

[illegible]

[illegible]

[illegible]

Challenges for Water Governance in Rural Water Supply: Lessons Learned from Tanzania

A. JIMÉNEZ* & A. PÉREZ-FOGUET**

*Ingeniería Sin Fronteras-Asociación para el Desarrollo (ISF-ApD), Madrid, Spain and Research Group in Development Co-operation and Human Development (GRECDH), Barcelona, Spain; **Research Group in Development Co-operation and Human Development (GRECDH), Barcelona, Spain

ABSTRACT *This paper focuses on the identification and analysis of key issues that impact the governance of rural water services in sub-Saharan Africa. Tanzania was selected as a representative case study. The analysis was based on a combination of relevant literature review, extensive fieldwork and action research case studies, which were carried out between 2005 and 2009. A number of weaknesses that continue undermining strategies for poverty eradication were identified at different administrative levels (from local to national): low quality of water services; lack of sustainability of constructed infrastructure; difficulties for targeting the poor; and inadequate internal information systems. Some initiatives to overcome these challenges were piloted and implemented at the district level. Policy recommendations presented entail new paradigms for the provision of rural water supply: adoption of water supply as a service that is monitored and supported by the government; needs-based allocation of projects at community level; and improving guidance for local government decision making are proposed.*

Introduction

It is estimated that 880 million people do not have access to safe drinking water and almost 2.6 billion have limited access to adequate sanitation (WHO & UNICEF, 2008). Technical or physical problems are rarely the reason for this appalling situation. To a large extent, these are socially and politically induced challenges, defined by the established water governance. This concept refers to the range of political, social, economic and administrative systems that are in place to develop and manage water resources, and to deliver water services to different levels of society (Rogers & Hall, 2003). In other words, water governance has been also defined as the system of actors, resources, mechanisms and processes which mediate society's access to water (Franks & Cleaver, 2007).

The UN Water Conference held in Mar del Plata, Argentina in 1977 proposed the period 1981–1990 as the International Water Supply and Sanitation Decade, with the aim of delivering water-related services for 100% of the world's population. Even though targets were not achieved, water and sanitation appeared for the first time as a top priority in the development agenda. This main concern was taken up once more during the last decade. The Millennium Development Goals (MDGs) include a specific target to cut in half, by 2015, the proportion of people that lack access to water and sanitation services. Furthermore, the decade 2005–2015 was declared International Decade for Action: Water for Life, but with very limited impact due to the lack of subsequent implementation plans. Along the same line, the year 2008 was declared the International Year of Sanitation, to stress the impact of poor sanitation and lack on hygiene on health, dignity and quality of life among millions of people.

Despite these political efforts, data show that there has been only moderate progress to date, and huge inequalities appear when comparing access to water with access to sanitation, rural with urban areas, and trends within different regions (Table 1). At the international level, monitoring of access to water and sanitation is being carried out by the World Health Organization (WHO) and the United Nations Children's Fund (UNICEF) Joint Monitoring Programme for Water Supply and Sanitation (JMP), and this provides a rough idea about the number of people with access to improved facilities. Nevertheless, these data are not exhaustive and present significant limitations (Jiménez *et al.*, 2009). In general, the paucity of consistent water sector related data is another key constraint that is impeding effective progress (Biswas & Seetharam, 2008).

In recent years there has been a significant change in how donors deliver aid. An effort is being made to improve aid effectiveness, as expressed in the Rome Declaration on Aid Harmonization in February 2003 and the Paris Declaration on Aid Effectiveness in March 2005. Improvements in aid effectiveness are based on the principles of ownership, alignment, harmonization, results-oriented management and mutual accountability. In practical terms, at least 85% of aid flows will be reported in government budgets and use public financial management systems. This ensures that most aid is channelled through sectorial or general budget support, which considerably increases ministries' budgets (Jiménez & Pérez-Foguet, 2009a).

Inside the sector, different positions have emerged. The Dublin Principles, which recognized water "as an economic good" (ICWE, 1992) gave support to the privatization of services. In contrast, a strong movement has defended the role of public institutions in the provision of basic social services (Hall & Lobina, 2004), and the wider principle of considering water as a common good (Barlow, 2009). However, the consideration of water as a human right contained in General Comment 15 of the Committee on Economic, Social and Cultural Rights of 2002 is a key milestone that enforces clear obligations on governments to protect, respect, and fulfil this right (United Nations, 2002). This right has not been accepted by many countries, since they are unsure of the legal implications of it (Biswas, 2007). Recently, the United Nations appointed an independent expert on human

Table 1. Access to water and sanitation and progress towards MDGs

Parameters	Water supply	Sanitation
World coverage	87%	62%
World rural coverage	78%	45%
World urban coverage	96%	79%
Estimated year for attainment of MDGs (world)	2016	2022
Estimated year for attainment of MDGs in Sub-Saharan Africa	2040	2076

Source: (UNDP, 2006)

rights obligations related to access to safe drinking water and sanitation to shed light on the matter.

The Situation in Rural Areas

Although the international private sector has focused its attention over the past decade on the urban water supply subsector, the Dublin Principles also address rural areas. They have been translated into what is known as the demand-response approach (DRA), which received considerable support during the 1990s (Katz & Sara, 1997). The underlying idea is that supply-led approaches are financially unsustainable and ultimately fail the poor. In contrast, a focus on water as an economic good and on the costs related to its supply lead to financial sustainability that results in improved services. Thus, users are brought into the process of selecting, implementing and ultimately financing the long-term delivery of water services. The main aspects of the DRA are summarized in Table 2. While this approach leads to greater participation of end-users in the design and management of their services, it also makes them liable for the responsibilities and costs related to full operation and maintenance (O&M). This approach has been widely accepted. In most cases, responsibility for the demand and management of rural supplies has been shifted to communities, and their support has been delegated to different local government agencies in many African countries (Politica Nacional de Aguas, 1995; Ministry of Water Resources, 1999; Directorate of Water Development, 1999; Ministry of Water and Livestock Development, 2002; Water Resources Commission, 2008).

However, the sustainability of rural water supply programmes remains a challenge. Current estimations for Sub-Saharan Africa suggest that only two out of three water points (WPs) in the continent's rural areas are functional at any given time. In Tanzania, a recent study estimates that 46% of public improved water points in rural areas do not work (Taylor, 2009a).

This paper analyzes how the general context in the sector affects the governance of rural water supply in Tanzania. In the next section, we describe a brief history of the development of water services in this country. This is followed by the methodology of this study. The Results section highlights the key challenges that the rural sector is facing, as well as the initiatives undertaken to overcome them. Policy implications are presented before the conclusions.

Water Services Development in Tanzania

Tanzania is a good example of the political processes that govern water service delivery models, due to its extensive experience in the development of rural water supplies. Such supplies were first established at the end of the 1940s, before independence, with 75%

Table 2. Main principles of the demand-response approach

Communities must initiate the process of making the demand, normally with initial financial contribution
Communities must contribute a certain percentage of capital costs towards their project (sometimes paid partially by in-kind labour) and 100% of O&M costs
Communities must participate in all decision-making steps
Communities own the system and are responsible for its management

funding from the national government and 25% from local authorities. Operation and maintenance costs were borne by local authorities by means of water rates and taxes.

In 1965, shortly after independence, the government decided that 100% of the funding for rural water supplies would be contributed by them, and that water at public domestic points would be free. However, maintenance was still the responsibility of local government. As local authorities were not effectively maintaining water points, in 1969 the government decided that it would also take on the responsibility for maintenance. Moreover, at the end of 1970, the ruling party, Tanganyika African National Union (TANU), created an ambitious plan which stated that by "the end of 1990, the people of rural Tanzania should have year round supplies of safe and wholesome water in sufficient quantities within a reasonable reach" (Tanzania Society, 1975). Lack of sustainability was already significant during the mid-1970s despite government promises.

As early as 1981, president Nyerere stated that the users, rather than the government, should look after the facilities, but this did not become policy until much later. The promises of rapid coverage were also fostered by the International Drinking Water Supply and Sanitation Decade (1981–1990). Donors, who provided 63% of the funds for the rural water sector in the country by 1980, widely supported this initiative and switched from programme to project aid, which mainly involved in-depth regional planning and support for further investment for implementation. In general terms, plans were not implemented and sustainability was very low (Therkildsen, 1988). The Decade began with a rural coverage of 46%, after government data, and ended with 39% coverage, after international estimations (WHO & UNICEF, 2009).

A new water policy was launched in 1991. A new target was established: to provide clean and safe water to the population within 400 meters of their households by the year 2002. This target was not achieved, since the rural coverage in 2000 was estimated at 44% (WHO & UNICEF, 2009). The main shortfall in the National Water Policy of 1991 was identified in the implementation strategies, which emphasized that the central government is the sole investor, implementer and manager of projects in rural and urban areas, while part of the O&M costs responsibility was shifted to the end-users.

In 2002, at the end of the target period, another water policy was launched, which is still valid today (Ministry of Water and Livestock, 2002). This policy was recently completed by the publication of the related act (Ministry of Water and Irrigation, 2009). In this act, water is recognized as a human right in the preamble, but no further implications of this statement are found throughout the document; the central government plays the role of coordinator and facilitator in the water sector, and the district level holds the main responsibilities for implementation. The approach to service delivery is the aforementioned demand-response approach, whereby: communities should demand, own, and maintain their water services and participate in their design; full operation and maintenance costs are their responsibility; and they have to provide part of the capital costs through cash and kind.

The main policy implementation instrument is the Water Sector Development Program, whose rural component is the Rural Water Supply and Sanitation Program (RWSSP). The RWSSP, which was officially launched in 2006, establishes targets for the percentage of the population in rural areas who shall have sustainable and equitable access to safe water: (1) 65% by 2010 (goal set by the National Strategy for Growth and Reduction of Poverty, MKUKUTA); (2) at least 74% by mid-2015 (MDGs); and (3) 90% by 2025. To meet these targets, water supply coverage must be extended to an additional 33.8 million people during the period 2005–2025. The estimated cost for the rural component (excluding small towns) is US$1,606 million, with US$1,465 million for capital investment, US$51 million for management and operational support to districts, and US$17 million for institutional strengthening and development (Ministry of Water and Irrigation, 2006).

Hence, Tanzania is a state that somehow recognizes the right to water, has decentralized competences, and takes a fully demand-responsive approach to service delivery. The current reported level of rural coverage is 57.1%, compared to the 65% targeted for 2010. The main challenges that compromise the success of the RWSSP are described hereunder, after an explanation of the methodology.

Methodology

The methodology used in this study combines quantitative data obtained from Water Point Mapping studies with qualitative data obtained through fieldwork and interviews, as well as an action research case study, which was carried out between 2006 and 2009 in one rural district, and involved some initiatives to overcome challenges.

Water Point Mapping (WPM) campaigns were undertaken in two districts, Kigoma Rural and Same, and covered 2509 water points and around 840,000 people. The campaigns were undertaken in 2006 and 2008 respectively. WPM can be defined as "an exercise whereby the geographical positions of all improved WPs in an area are gathered in addition to management, technical, and demographical [*sic*] information. This information is collected using GPS and a questionnaire located at each WP. The data are entered into a geographical information system and then correlated with available demographic, administrative, and physical data. The information is displayed using digital maps" (Welle, 2005). These studies covered the standard water point mapping information plus basic quality testing and seasonality assessment (Jiménez & Pérez-Foguet, 2008, 2009b), named as Enhanced Water Point Mapping.

Additionally, this information was completed with an analysis of the information obtained from other WPM studies undertaken between 2005 and 2006 that covered three regions and 15 districts of central Tanzania (Jiménez & Pérez-Foguet, 2009c). These studies involved 5921 water points and around 15% of the country's total rural population in 2005.

Qualitative information about the RWSSP at the national level was obtained through interviews with Ministry of Water officials, along with an extensive review of the unpublished and published documents from this ministry and the office of the prime minister. The analysis of the main challenges at the decentralized level was based on fieldwork conducted in four rural districts (Kigoma rural, Same, Iramba and Nzega) between July 2008 and August 2009. District councils were visited and interviews were held, particularly with district water engineers (DWEs) and district planning officers (DPLOs). For the purpose of understanding the challenges at lower levels of government, four wards were selected in two districts (Same and Nzega). The aim was to include one ward with a historically low investment in water supply and one with a historically high investment in each of the districts. Interviews and group discussions were held with elected political representatives at ward, village and sub-village levels, as well as with government officers at village and ward levels. The village plans from the selected wards were examined and discussed with government officials and local leaders.

The action research case study was carried out in Same District between 2006 and 2009 through collaboration between Same District Council and Ingeniería Sin Fronteras, a Spanish non-governmental organization (NGO). An initial WPM study was conducted at the end of 2006 as a baseline on the state of water services in Same District. The application of the tool was monitored with a focus on the coordination of stakeholders and decision-making regarding the water funds allocation of the District Government. In July 2009, a basic WPM update was carried out to assess its evolution from 2007 to 2009. This was used to test a pilot initiative on how to update the WPM and to link the available information with the planning. This process led to the definition of initiatives to overcome the challenges, as described in Section 5.

Challenges for Water Governance in Rural Areas

Low Quality of Delivered Service

The aim of the Rural Water Supply and Sanitation Program (RWSSP) is to provide safe and sustainable water services to the rural population. However, an in-depth analysis of current services shows that there are major threats to this target. The quality of water delivered, and the reliability of the supply were analyzed through the Enhanced Water Point Mapping methodology, in two districts (Jiménez & Pérez-Foguet, 2009b).

The quality of water delivered was not satisfactory, especially due to coliforms. When the information was disaggregated by category, about 40% of the ground water points were found to be polluted, together with 30% of gravity-fed systems. Seasonality also affected the services in up to 30% of cases, depending on the category and geographical location of the water point. When we assimilated the results to their corresponding networks, coverage was reduced by one quarter when the presence of coliforms was considered, and by 20–33% with seasonality. When both quality and seasonality were combined, coverage figures dropped by a factor of 0.57 and 0.55 for the districts, compared to the coverage figures that reflect functionality alone.

The study shows that more than 50% of functional improved water points could be expected to have either quality or seasonality problems, which is in agreement with similar studies in the literature. Unpublished results from the Rapid Assessment Quality Water pilot test in Ethiopia were similar. Of the 290 boreholes tested, 23.1% had more than 10 CFU/100 ml, compared to 34.2% of the 155 protected dug wells and 46.7% of the 319 protected springs.

There are various explanations for this fact. However, in the case studied, many of them are related to poor management of services rather than infrastructure failure or natural sources of pollution. As regards quality, contamination at source was predominantly due to bad management of water catchment. In a few cases, water was naturally polluted (salinity or fluoride) since this aspect is increasingly controlled when new water points are created. Seasonality was related to depletion and bad use of sources due to (1) inappropriate land uses around the source, (2) poor allocation to different uses of water abstracted from the same source, and (3) uncontrolled connections to the network, which produce shortages in the dry season.

This situation clashes with the current plans at the national level. On the one hand, quality monitoring and risk assessments are not part of national information routines. Moreover, the lack of capacity at the community level to deal with the mix of environmental and social aspects that affect the quality and seasonality of the water consumed, contrasts with the scant attention paid to conflicts about water use, capacity building and post-project support foreseen in the RWSSP.

Low Sustainability

The RWSSP emphasizes the development of new schemes. It allocates less than 6% of investments for rehabilitation and less than 4% for district management support and capacity building. This allocation of resources is challenged by the study of current water point functionality-time relationships we made for a water point mapping survey conducted in three regions of Tanzania that account for 15% of the country's total rural population (Jiménez & Pérez-Foguet, 2009c). In this study, functionality was disaggregated by technology category and administrative structure, and appropriate scales of analysis of the various relationships were justified. The functionality by category showed that only 45.3% of hand pumps, 48.6% of gravity-fed systems and 44.4% of motorized systems were functional at the time of the survey. Some WP categories were found to be quite sustainable in some areas and to fail completely in others. Nevertheless, the analysis showed a statistically significant relationship between functionality and category of WP at all administrative levels, national, regional and, to a lesser extend, district ones.

Decreasing functionality rates over time were found for all WP categories. If we analyze the tendencies showed by the lineal regression we find that hand pumps show the least favourable functionality-time function, starting from 61% in the first five years to 8% in the 30-year period. Motorized systems decrease from 79% to 17% in the same period, and gravity fed systems, from 67% to 19%, working better in the long run that any other category of water point. In all categories, only between 35% and 47% of water points are expected to be working 15 years after installation, and only between 57% and 72% of them continued working after five years. The latest data about the implementation of the pilot phase of the RWSSP (2002–2008) confirm the conclusions of this analysis (World Bank, 2008). Out of 197 water points examined in 19 sampled systems that were implemented in 6 districts during the last five years, 130 (66%) were functional at the time of the evaluation, with a 75% functionality rate for gravity and 56% for hand pumps. These values show that the functionality-time tendency has not changed with the current implementation model.

RWSSP predictions estimated that 48% of people will be served by hand pumps, 25% will be served by motorized systems and 21% will be served by gravity-flow networks. Hence, the level of service provided and the technology proposed for the rural areas need to be reviewed, as the most predominant technology, the hand pump, is the least sustainable over time. It is true that the community management requirements are lower than any other: this in turn might have led to the scant attention paid to building organizational capacities, which remain critical for maintenance. Moreover, it was expected that people would prefer the low cost-low service option. However, this is not the case, as the evaluation of the first phase of the RWSSP shows (World Bank, 2008).

Sustainability is threatened by the limitations of community management of funds, the difficult relationship between water user entities and elected village representatives, the low professionalism in the management of services (Giné & Pérez-Foguet, 2008), and the very limited role that decentralized government plays with regard to monitoring, regulation and technical support, among other factors. The policy and the RWSSP are vague in defining the setups that are possible at the community level to manage the service effectively. The main responsibility is given to the community, but much greater support needs to be provided to attain effective sustainable service management models.

Lack of Pro-poor Targeting

The allocation of funds at the ministry level under RWSSP is a fairly transparent formula-based system. However, a thorough study reveals some drawbacks. First, it is too focused on the development of new infrastructures (91.2% of budget of the programme) and low priority is given to capacity building and post-project support. In fact, recurrent costs are not included in the budget of the programme, and depend on transfers from the Ministry of Local Government (PMO-RALG). Second, it is focused on efficiency rather than on regional equity, despite the initial goal to raise coverage in all districts to over 80% by 2025. The main driver of allocations in practical terms is the total number of people with no water service in a district, with a minor influence of technology. In fact, the allocation data analyzed versus the population show that there is a good relationship between the overall number of people with no water service in a district and the money allocated ($R^2 = 0.95$), but not between the money allocated and the coverage rate by district ($R^2 = 0.21$). Third, differences between formula predictions and real allocations are important. Some regions get significantly more funds than the water formula would allow for, while in other regions the opposite holds true (World Bank, 2009).

Nevertheless, the greatest challenges for targeting the poor are found at district level. District councils allocate projects based on a combination of need, demands (expressed in cash) and political influence. This tends to help bigger villages that are better connected and more influential. Thus, existing inequalities are perpetuated. This situation is not counterbalanced by regular awareness creation and facilitation in villages that are less organized or have worse connections. The dynamics of these districts are unlikely to change in the short term from the bottom level. Villages and councillors are not sufficiently aware of programmes other than LGG, and only selected villages are being supported by the RWSSP to complete their applications and initial contributions. In addition, villagers are ill-informed of application procedures and decision-making processes.

This mixture of policy incoherencies, technical shortcomings and political influence determines that only a small proportion of funds reach the under-served areas. A close look at four districts showed that, apart from the above mentioned facts, only 50% of the wards that were targeted in the first phase of RWSSP in these districts were below the corresponding district average of access. This is a common problem, as experience has shown that when governance is decentralized, local elite are frequently even less likely than the national elite to target government resources to the poor (Blair, 2000). Simultaneously, people's capacity to participate and hold local government accountable is reduced, especially for the poorest (Cleaver, 2005).

Inadequate Information Systems

The failure to target the poor is also due to the lack of suitable and reliable information systems that show the status of water services across the region. This is a general concern for the sector, as demonstrated by the status of development of Sector Management Information Systems in the Sub-Saharan region (Water and Sanitation Program, 2007).

The case of Tanzania reflects common challenges. Data published by the ministry, which are based on the coverage reported by districts, are not always reliable. District water engineers recognize that data are not based on an extensive review of the situation.

Inter-annual variability is also very high. For instance, from 2007 to 2008, 30 districts reported a coverage variation of at least 10% on the previous year. Of these, 16 reported a variation of over 20%, and seven reported one of over 30% (Ministry of Water and Irrigation, 2008).

This aspect had been identified before, and it has been tackled by promoting Water Point Mapping, which has been supported by international NGOs since 2005. So far, 51 districts out of 132 have been mapped, and the government might extend this scheme to the entire country. This exercise has shown a much more reliable picture of the status of water inside the districts, and has highlighted major differences between official and onsite data, as well as significant internal inequalities.

Despite the use of WPM as an information tool, its potential remains underexploited. A field study was carried out to assess the use of WPM in four districts in which it had been in place since 2005 (Taylor, 2009b). The results showed that the incidence of WPM for better planning was still low, despite the acknowledgment of its potential usefulness. The main constraints were related to updating the system and how it can be effectively included in the planning process.

Initiatives Implemented

Some initiatives have been tested and proposed to overcome the main challenges detected. These are described below.

Promotion of Enhanced Water Point Mapping as an Information Tool

As it has been said, lack of reliable information is at the heart of some of the main problems of the sector. In this sense, Water Point Mapping (WPM) was created to overcome some of these difficulties and has been widely used in the country, but is facing the challenges for its updating as well as for the effective use of information. Two initiatives were tested.

- Implementation of a yearly basic update of WPM (Jiménez & Pérez-Foguet, 2010a). Currently, a typical district council does not have enough resources or capacity to repeat a whole WPM campaign every year. Consequently, a simplified procedure was tested in Same District which reduced costs and the need of highly qualified staff. Instead of visiting each water point, information was collected at the village level through meetings with local leaders. Thus, information on all water points that had already been recorded in the initial WPM database was updated and completed with a list of new WP. This basic update of WPM gathers information about water points summarized by village. The position of new water points is not recorded with GPS, but the name and location of up to sub-village level is available, and maps can be produced. This basic update is not intended to substitute a whole campaign to be done every four to five years, but helps to develop a reliable information system at the local level in the following aspects: (1) it gives an updating option more adapted to the current resources and capacities of the districts; (2) it involves district officials and local leaders in the collection of information; (3) it provides reliable information that can be used for planning at district level, as described hereinafter.

- Inclusion of basic quality testing and seasonality information in the WPM campaign. The Enhanced Water Point Mapping (EWPM) was piloted in two Districts (Jiménez & Pérez Foguet, 2009b), and as described above, unmasked important problems related both to quality and seasonality of the services. This facilitates the adoption of some corrective activities from the district level, as described hereinafter.

Link of Water Point Mapping Results to District Planning

WPM offers great potential in terms of analysis and planning, but it remains underexploited to date. For the pilot action implemented together with the Same District Council, WPM was included in a wider framework for improving planning, which included three main actions: (1) priorities based on objective data were defined using the results of the WPM update; (2) demand creation at community level has to be included in the local government authority's (LGA's) activities, to prevent funds from being allocated only to the most prepared and organized villages; and (3) information systems had to be regularly updated to feed the process, as described above. The aim of the process is that the LGAs focus on supporting underserved communities to cope with policy requirements, and finally provide services where they are most needed. Priorities were defined on the basis of need, with regional equity as the key driver, in order to achieve universal coverage. For this purpose, some basic indices were defined using the information obtained from WPM (Jiménez & Pérez-Foguet, 2010b). Planned actions included an increase of coverage (through new water points and rehabilitation), an increase of equity at the village level (by targeting sub-villages with no access), improvement in the quality of water delivered and the implementation of environmental actions (particularly those related to source protection). Moreover, basic information on service management at the village level helped to identify which villages must be more urgently supported to establish suitable management systems at the community level. The output of the process was a district-owned planning based on need, rather than demand, as the main criterion (Same District Council, 2009a), and included priorities for a wider range of activities, as defined above.

Establishment of District Regular Management Support Services

As described, low sustainability remains the greatest challenge in the Tanzanian rural water supply. This is a wide and complex issue that has various causes. Nevertheless, one of the main weaknesses is the absence of an institutional arrangement at the district level to provide long-term support to community-managed water services. Community rural supplies need to be monitored and supported regularly by the appropriate level of government, the LGA in the Tanzanian case. However, some challenges will need to be overcome: (1) the funds for recurrent costs at the LGA level remain very low, which makes it difficult to effectively support O&M at community level; (2) the different aspects that threaten sustainability are above the capabilities of Water Departments alone; (3) LGAs lack human resources in many areas.

A proposal was developed together with Same District Council, named as the District Water and Sanitation Unit Support (DWUS). The DWUS has been designed as a multisectoral team comprised of members from nine departments: Water, Health,

Education, Community Development, Finance, Planning, Forestry, Land, and Legal issues, in order to assist communities in the different challenges that may arise. The expected outcome is an increase in the sustainability rates of the rural water and sanitation services, through the establishment, legalization and support of water user entities (Same District Council, 2009b). The operational rules of DWUS have been developed taking into account the aforementioned current limitations of the LGAs.

Policy Implications

This research explores possibilities to improve water-related services delivery in rural areas. Achieved results pointed out in previous section support the following policy recommendations.

Sector information systems are more useful when (1) data collection involves end users and promotes a thorough description of reality on the ground (through an adequate mix of survey instruments), (2) they produce outputs that are not only valuable for reporting to higher levels of government, but can also be used for decision making at intermediate levels, (3) the information is linked to territory, and thus it might be displayed via digital maps to facilitate interpretation and analysis; (4) updating can be done in the short term relying on available capacities at local level.

The provision of improved water points is not enough to ensure safe drinking water quality. Thus, the inclusion of basic quality parameters and seasonality monitoring in the rural supplies is required to unmask important shortcomings in service provision, which might have undesirable effects on the well being of millions of rural users. Moreover, the expected costs for considering these aspects are relatively low, while costs for standard mapping range from 12–15 US$/WP, they increase to 20 US$/WP when quality is included. In Tanzania, if enhanced water point mapping is applied to the entire country, the total cost would be roughly US$2 million compared to US$950 million of foreseen investment throughout the programme for the period 2008–2012.

The design of national water plans should include the necessary institutional arrangements and funds to provide LGAs with adequate resources and capacities to supervise works and ensure post-project support to community water supplies. As a current constraint, this is particularly grave since it does not represent an important amount of funds when compared with investment for infrastructure, but with enormous consequences in the sustainability of services.

The national plans have to ensure an adequate channelling of funds from the ministry level to the end users. In this aspect, increased decentralization of responsibilities can prevent funds from reaching the neediest, in those contexts with weak democratic processes at the local level. Hence some measures are proposed. First, equity in service provision at the local level should become an explicit target to be monitored from central governments. National directives should be in place to guarantee a minimum level of service per ward and village, as is done for other social services. Second, national plans should include in periodic evaluations some performance indicators related to equity and functionality rates at the decentralized level, with incentives to well performing districts. Third, transparent mechanisms need to be developed to link monitoring information with decision making at the local level.

National policies often state that community rural services have to be allocated through a demand-response approach. An adequate interpretation of 'demand' is far from easy. This has frequently been measured in terms of the amount in cash that a community is able to collect, which facilitates the influence of local political powers. Needs-based allocation of resources should be a must in rural water policies. Demand creation and facilitation should be effectively included in the project cycle, but not as a pre-requisite.

Conclusions

The aim of this work was to study some of the key governance issues that affect the water sector in developing countries, especially in the rural areas of Sub-Saharan Africa.

Tanzania was selected as case study. A comprehensive analysis of main challenges was undertaken, and some initiatives were piloted to improve policy making towards better service delivery. The research, which addressed levels from national to village, allowed some conclusions to be drawn.

The rural sector is dominated by important investment plans to increase access within more global strategies for poverty eradication. These usually occur together with a decentralization process. Service delivery in the rural context relies on a demand-response approach at community level.

The overall performance is constrained by important weaknesses. Low quality and sustainability of the service, lack of pro-poor targeting, and inadequate information systems were found to be the most significant challenges. Important changes in policy orientation need to be addressed to improve the performance of sectors. These involve the shift of some paradigms.

First, national policies and plans need to change from an infrastructure to a service approach. This implies that adequate resources should be envisaged for the operation and management of services in the medium term. Allocation of responsibilities for the management of the services has to be redefined. A greater balance between the participation of end users in the management of services and an adequate support and control from government institutions needs to be achieved. In this sense, the role of local governments has to be strengthened. They should effectively monitor and regulate service and provide technical support. Different possible setups (management by the community itself, outsourcing of some tasks to private service providers, total warranty schemes, and so on, have to be developed at the community level to enable management of the service.

Second, and in terms of equity, project allocation decisions cannot be based on the demand of communities. Plans should be based on real needs, so that unorganized, poor and small communities are not side-stepped from service delivery, and universal coverage can be achieved.

Third, decentralization is not beneficial for citizens *per se.* It is both a risk and an opportunity. Clearer orientations and incentives from central governments could be useful in the short term, together with the development of procedures to link available information to political decisions. Meanwhile, accountability to citizens and transparency has to be dramatically increased so that decentralization leads to better performance. On the other hand, the initiatives implemented in this research show that improvements can be easily tested and adopted at the decentralized level, and that it is worth supporting these institutions.

Finally, the establishment of reliable and inclusive information routines for sectors is the key ingredient for many of these changes to be possible, as well as to anticipate future challenges. Objective, reliable and detailed information about water access is essential.

Tools based on GIS, like the Water Point Mapping, have great potential. But above all, the will of having reliable monitoring systems in the water sector should become a real priority for international donors and governments.

Acknowledgements

The authors would like to thank Ingeniería Sin Fronteras-ApD (ISF-ApD), the Civil Engineering School and the Centre for Cooperation and Development (CCD) of the Universitat Politècnica de Catalunya (UPC) for supporting this research. The authors would like to express as well their sincere gratitude to Same District Council for their enthusiastic collaboration during these years.

References

Barlow, M. (2009) *Our Water Commons: Towards a New Freshwater Narrative* (West Ottawa, ON: The Council of Canadians). Available at www.canadians.org/water/publications/water%20commons (accessed January 2010).

Biswas, A. K. (2007) Water as a human right in the MENA region: challenges and opportunities, *International Journal of Water Resources Development*, 23(2), pp. 209–225.

Biswas, A. K. & Seetharam, K. E. (2008) Achieving water security for Asia, *International Journal of Water Resources Development*, 24(1), pp. 145–176.

Blair, H. (2000) Participation and accountability at the periphery: democratic local governance in six countries, *World Development*, 28(1), pp. 21–39.

Cleaver, F. (2005) The inequality of social capital and the reproduction of chronic poverty, *World Development*, 33(6), pp. 893–906.

Directorate of Water Development (1999) *National Water Policy* (Kampala, Uganda: Directorate of Water Development).

Francis, P. & James, R. (2003) Balancing rural poverty reduction and citizen participation: the contradictions of Uganda's decentralization program, *World Development*, 31(2), pp. 325–337.

Franks, T. & Cleaver, F. (2007) Water governance and poverty: a framework for analysis, *Progress in Development Studies*, 7(4), pp. 291–306.

Giné, R. & Pérez-Foguet, A. (2008) Sustainability assessment of national rural water supply program in Tanzania, *Natural Resources Forum*, 32(4), pp. 327–342.

Hall, D. & Lobina, E. (2004) Private and public interests in water and energy, *Natural Resources Forum*, 28, pp. 268–277.

International Conference on Water and the Environment (1992) The Dublin Statement of Water and Sustainable Development. Available at http://www.wmo.ch/pages/prog/hwrp/documents/english/icwedece.html (accessed January 2010).

Jiménez, A., Molinero, J. & Pérez-Foguet, A. (2009) Monitoring water poverty: a vision from development practitioners, in: M. R. Llamas, L. Martinez Cortina & A. Mukherji (Eds) *Water Ethics: Marcelino Botín Water Forum 2007* (London, UK: Taylor and Francis).

Jiménez, A. & Pérez-Foguet, A. (2008) Improving water access indicators in developing countries: a proposal using water point mapping methodology, *Water Science & Technology: Water Supply*, 8(3), pp. 279–287.

Jiménez, A. & Pérez-Foguet, A. (2009a) International investments in the water sector. *International Journal of Water Resources Development*, 25(1), pp. 1–14.

Jiménez, A. & Pérez-Foguet, A. (2009b) Access to safe and year round functional water: an estimation of coverage for three central regions in Tanzania. Proceedings of the 34th WEDC International Conference, Addis Ababa, Ethiopia.

Jiménez, A. & Pérez-Foguet, A. (2009c) Consequences of low sustainability in the effectiveness of national strategies to increase water access in the rural areas: evidence from three central regions of Tanzania. Proceedings of the 1st IWA Development Congress, México, 2009.

Jiménez, A. & Pérez-Foguet, A. (2010a) Targeting the needs in the water sector: the potential of water point mapping—a case study from Tanzania. *Proceedings of the Biennial Conference of the Water Institute of Southern Africa*, 18–22 April, Durban, South Africa.

Jiménez, A. & Pérez-Foguet, A. (2010b) Building the role of local government authorities towards the achievement of the right to water in rural Tanzania. *Natural Resources Forum* (in press).

Katz, T. & Sara, J. (1997) *Making Rural Water Supply Sustainable: Recommendations from a Global Study*. Report for the Water and Sanitation Program, UNDP and World Bank. Available at http://www.its.caltech.edu/~e105/readings/rural_water.pdf (accessed January 2010).

Ministry of Water and Irrigation (2006) *Water Sector Development Program 2006–2025: Main Report* (Dar es Salaam, Tanzania: Ministry of Water and Irrigation).

Ministry of Water and Irrigation (2008) *Water Sector Performance Report for the Year 2007/2008* (Dar es Salaam: Ministry of Water and Irrigation).

Ministry of Water and Irrigation (2009) *The Water Supply and Sanitation Act*, No. 12, 15 May 2009.

Ministry of Water and Livestock Development (2002) *National Water Policy* (Dar es Salaam, Tanzania: Ministry of Water and Livestock Development).

Ministry of Water Resources (1999) *Ethiopian Water Resources Management Policy* (Addis Ababa, Ethiopia: Ministry of Water Resources).

Politica Nacional de Aguas (1995) *Boletim da Republica* (Publicação Oficial Da República De Moçambique) 23 August 1995.

Rogers, P. & Hall, A. W. (2003) *Effective Water Governance*. Global Water Partnership Technical Committee (TEC) Background Papers No. 7 (Stockholm: Global Water Partnership). Available at: http://www.gwpforum.org/gwp/library/TEC%207.pdf (accessed 11 April 2010).

Same District Council (2009a) *New Water Supply and Sanitation Services Priorities in Same District*, September 2009. Prepared by Same District Water Engineer, P.O. Box 7, Same, Tanzania.

Same District Council (2009b) *District Water and Sanitation Unit Support* July 2009. Prepared by Same District Water Engineer, P.O. Box 7, Same, Tanzania.

Tanzania Society (1975) Water supply; service to the nation to speed development. Tanzania notes and records, 76 (Dar es Salaam: Tanzania Society).

Taylor, B. (2009a) *Addressing the Sustainability Crisis: Lessons from Research on Managing Rural Water Projects* (London: Overseas Development Institute). Available at: http://www.wateraid.org/international/what_we_do/policy_and_research/4602.asp?Keywords=&Subject=&Author=&Country=16&Language=0&Sort=Date&Display=results (accessed 11 April 2010).

Taylor, B. (2009b) *Waterpoint Mapping, Planning and Obstacles to Equity in Rural Water Supply: A Review in Mpwapwa, Kongwa, Iramba and Nzega* (Dar es Salaam: WaterAid Tanzania).

Therkildsen, O. (1988) *Watering White Elephants? Lessons from Donors Funded Planning and Implementation of Rural Water Supplies in Tanzania* (Uppsala: Scandinavian Institute of African Studies).

United Nations (2002) *The Right to Water*, Economic and Social Council Committee on Economic, Social and Cultural Rights, E/C.12/2002/11, General Comment No. 15, 26 November 2002.

United Nations Development Programme (2006) *Human Development Report, 2006: Beyond Scarcity—Power, Poverty and the Global Water Crisis* (New York, NY: UNDP).

Water and Sanitation Program (2007) Country-level sector information and monitoring systems (SIMS) for water and sanitation in Africa. Practitioners Workshop, Nairobi, Kenya, March 2007. Available at http://www.wsp.org (accessed January 2010).

Water Resources Commission (2008) *National Water Policy* (Accra, Ghana: Water Resources Commission).

Welle, K. (2005) *Learning for Advocacy and Good Practice: WaterAid Water Point Mapping* (London: Overseas Development Institute). Available at: http://www.wateraid.org/international/what_we_do/policy_and_research/4602.asp?Keywords=429&Subject=&Author=&Country=0&Language=0&Sort= Date&-Display=results (accessed 11 April 2010).

WHO & UNICEF (2008) *Progress on Drinking Water and Sanitation: Special Focus on Sanitation Joint Monitoring Programme for Water Supply and Sanitation* (Geneva: WHO). Available at http://www.who.int/water_sanitation_health/monitoring/en/ (accessed January 2010).

WHO & UNICEF (2009) Tanzania data summary sheet. Available at: http://mdgs.un.org/unsd/mdg/SeriesDetail.aspx?srid=667 (accessed 11 April 2010).

World Bank (2008) *Implementation Completion and Results Report (IDA-36230, IDA-3623A) on a Credit in the Amount of SDR20.8 Million (US$26.00 Million equivalent) to the United Republic of Tanzania for a Rural Water Supply and Sanitation Project*. Report No: ICR 0000730, Washington, DC: World Bank.

World Bank (2009) Tanzania Public Expenditure Review of the Water Sector. September 2009. Available at: http://www.wds.worldbank.org/external/default/main?pagePK=64193027&piPK=64187937&theSitePK=523679&menuPK=64187510&searchMenuPK=64187283&theSitePK=523679&entityID=000333037_20091006234617&searchMenuPK=64187283&theSitePK=523679 (accessed 11 April 2010).

Governing to Grow Enough Food without Enough Water—Second Best Solutions Show the Way

DAVID MOLDEN*, JONATHAN LAUTZE*, TUSHAAR SHAH**, DONG BIN†, MARK GIORDANO* & LUKE SANFORD*

*International Water Management Institute, Pelawatte, Battaramulla, Colombo, Sri Lanka; **International Water Management Institute, Anand, Gujarat, India; †State Key Laboratory of Water Resources and Hydropower Engineering Science, Wuhan University, PR China

ABSTRACT *As economies develop and societies change, emerging sets of challenges are placed on water resources and its governance. Population growth and economic development tend to drive the demand for more water, and push river basins into situations of scarcity. Agriculture, globally the largest user of water, is a major driver of water scarcity, and also the sector that has to bear the consequences of scarcity. Yet governance arrangements the world over have difficulty coming to grips with the management of agricultural water within the larger water resource context. The four major agricultural water governance challenges are: to manage transitions from abundance to scarcity; to deal with the large informal sectors of the agricultural water economy; to adapt to the changing objectives of society; and within each of these challenges, to craft context-specific solutions. This paper presents examples of these challenges and uses them to derive a conceptual framework to help us understand present agricultural water-use contexts, and to develop context specific solutions. The framework is based on two important and shifting contextual dimensions: the degree of scarcity within a basin, and the degree of formality in water use. Looking at agricultural water governance within this framework shows that some standard prescriptions for water problems may not always be appropriate and that 'second best' solutions can in fact be the best way forward. The challenge for governance is to facilitate the development of these solutions.*

Introduction

Change is inherent to water–human interactions, yet water governance and management structures are rarely set up to cope with transition. One dimension of change comes with development of water resources, where once seemingly abundant supplies are now insufficient to meet demand, posing new sets of issues to water managers. Change drivers outside the water sector also have an impact on water resources including competition between sectors, changes in a country's economy and governance structures, and climate change. Often accompanying these changes is a shift from informal to formal water economies.

While water for cities, energy, environment, and the integration across sectors will be important, the real 'elephants in the room' are the transitions in agricultural water management. Looking to the future, it is likely that nearly twice as much water will be consumed by agricultural uses by 2050, depending mainly on population growth, diet and production practices (Molden *et al.*, 2007b). Physical water scarcity is already a reality in major breadbaskets across the world, including the North China Plain, the Indus river, the Colorado River, and the Murray Darling in Australia. Agricultural water use will drive increased competition across sectors, and if not adequately managed, will cause inequities and further environmental degradation. While the paper has lessons for water governance in general, the focus is on agricultural water management, which requires relatively more attention to resolve issues of scarcity ahead of us.

This paper reviews four key challenges for governance in the water and agriculture sector in order to develop insights that broaden the conventional framework for understanding and responding to change in agricultural water management. The first challenge is to manage transitions as river basins move from a state of water abundance to water scarcity due to intensive development of water for agriculture. The challenge for water governance is to promote management structures that can quickly adapt to these changes. The second is recognition of the informal nature of much of water use for food. In many countries, a multitude of irrigators are self-suppliers and outside the range of government regulations. The governance challenge is to figure out how to engage these users to provide better service, and also to regulate use so that the impact on water resources is minimized. A third challenge is for agriculture to adapt to the multiple and changing objectives of society. At present, there is a growing trend to reallocate water from agriculture to cities. The governance challenge is to find ways to reallocate water and its related benefits and costs in a manner that promotes productive agriculture, yet meets the growing demands of cities and environments. A fourth challenge is that the development and management of water resources in some developing countries is heavily influenced by norms produced in more developed countries and in different contexts. Such norms are often not suitable for the conditions in the countries where they are implemented, and assume either a more formal system of governance, or more institutional capacity than is present. Furthermore, norms surrounding energy, environment and changes in dietary patterns have profound influence water use. The governance challenge is to craft solutions, cognizant of these norms, but practical enough to respond to local realities.

Based on these experiences, this paper lays out a new conceptual framework for understanding transitions in agricultural water management. The framework considers the conditions of most river basins as falling into one of four quadrants in a matrix where values trend from open to closed on one axis, and informal to formal on the other. Considering the experiences of several basins, the paper finds that there is no single path that basins follow, nor is there a best path to follow. Indeed, the trajectory of some basins has proceeded from conditions of open–informal to closed–formal to closed–informal institutions, while trajectories of other basins have gone from open–informal to closed–informal to closed–formal institutions. These findings help clarify the different permutations of physical and institutional water resources change, and help reveal a set of 'second best practices' which may be useful to countries responding to conditions in different contexts.

From Open to Closed Basins

A typical water-use pattern emerges in river basins where as society grows and prospers, more water is withdrawn and depleted to meet demands for cities, industries and agriculture (Keller *et al.*, 1998; Molden *et al.*, 2005; Molle, 2003; Falkenmark & Molden, 2008). When water is abundant the most common response to meet demand is to harness more water supplies. There comes a point where nearly all of the water resources have been fully developed, and adding more supplies from within the basin is no longer an option. This transition from open basins (Seckler, 1996), where adding supplies is an option, to closed basins, where water is fully or over-allocated, has happened over a relatively short time frame of 10 to 30 years in several key river basins with the development of large infrastructure, including the Yellow River in China, the Indus River in Pakistan, the Nile River in Egypt, and the Bhavani River in India. In an open basin there are more degrees of freedom for development, because adding supply is an option. In a closed basin, options are limited, and consideration is mainly given to demand management responses. A common pitfall in closed basins is when some people develop more blue water supplies in one area at the expense of someone else in the basin (Molle *et al.*, 2004). If a basin goes past the point of closure, the basin is in a danger zone, and a question of sustainability arises, because it is not possible for river basin water to adequately support its many functions, and ecosystem services will be lost unless societies figure out a way to curtail water use or increase supply. Nonetheless, a well-managed closed basin can sustainably support agriculture, urban and other ecosystem services, especially if environmental flows are judiciously defined and managed.

More than simply good management, this type of change could benefit from good water governance. Despite ambiguity surrounding the concept, there appears some convergence toward considering water governance as how people make decisions about how to manage water. In other words, whereas water management may include more mundane issues, such as operating reservoir releases to achieve efficiency and environmental sustainability, water governance includes broader aspects of how these decisions are made, such as how much public participation, transparency or integrity was involved in decisions. The concept differs from that of management; where management controls such things as irrigation operations and preventing pollution, governance includes *how* water should be shared between uses and regions, or *how* the standards for water quality are set and enforced (Tropp, 2007; Iza & Stein, 2009). While good management helps produce good water outcomes, it seems difficult to define what constitutes good water outcomes in a particular setting without the participation and transparency associated with good governance. In the face of the many water transitions taking place, good governance can help ensure that risks and rewards are consistent with societal preferences.

As river basins move toward closure, a transition in management is required as indicated in Table 1. This clearly shows the need for flexibility and adaptive management. Institutions tend to have difficulties in making this transition, as there is a natural resistance to change. This institutional rigidity will be the biggest impediment to the required change needed to address tomorrow's world.

In early stages of development, water governance and management structures are set up to exploit water resources. Main tasks are to mobilize money and to build infrastructure to provide new allocations of water to people (Hooper, 2006). Who gets the benefit from water, and who does not, are important political decisions. In open basins, it is possible

Table 1. Institutional changes as a basin makes the transitions from open to closed

Transition from	To
Exploiting water resources	Managing demand
New allocations	Reallocating water
Who is included or excluded from benefits	Safeguarding rights to water
Developing groundwater	Regulating groundwater
Dilution as a way to manage pollution	Control pollution at source
Single sector management	Coordinated management across sectors
Within country management	Transboundary arrangements
Accepting some environmental impact	Valuing and managing for ecosystem services
Manage per design	Deal with uncertainty, manage for change

to tap into groundwater without adversely impacting others. As a basin evolves from open to closed, new infrastructure will not necessarily provide new allocations to new people because there is simply not enough water in river basins. The result of new infrastructure is to tap into ecological reserves, or lower groundwater reserves. The governance challenge when basins transition from open to closed is then how to regulate water use such that it is not stretched beyond limits. In closed basins, new water allocations typically result in reallocations of water. Water rights become increasingly discussed as people wish to safeguard their claims to water. This is particularly difficult if water rights have not already been fixed, because it is difficult to go back in time to establish use patterns. People with more influence will tend to have a much more powerful voice in obtaining water rights than poor rural farmers. For rivers crossing international boundaries, reallocation becomes an international issue, calling for transboundary agreements (Giordano *et al.*, 2008). Water rights reallocation could be part of the solution in a closed international basin, however renegotiation of these rights becomes an issue of difficult international politics.

With the transition from open to closed, a complex flow web is developed, and it is quite difficult to see how one use impacts another. Groundwater is not seen and rarely adequately assessed, yet more pumping by one individual may easily impact another. It is in fact difficult to find cases where groundwater is adequately regulated (Shah *et al.*, 2007; Giordano & Vilholth, 2007). Similarly other issues emerge during the transition. In open basins it may be entirely appropriate to develop and manage water for a single sector without much interest in other sectors. Water developers tend to use the argument that the impact of a single structure is small, and that societies must bear some negative ecological consequences. With plentiful water, dilution is often viewed as an answer to pollution problems. With closed basins, there is no more water left to dilute polluted water. The sum of massive development tends to have large and unexpected consequences, and with competition for water an integrated view is warranted.

There have been significant drivers of change at play with irrigation that go far beyond the open/closed basin concept presented here. Shifts from rural to urban economies, massive changes in cropping patterns, more groundwater use, a 'new rurality' characterized by people with multiple sources of income outside of agriculture, an aging population doing agriculture, and shifts in gender roles—have all made irrigation much different than it was 20 or 30 years ago (Rauch, 2009). Yet with new irrigation, we aimed at managing per design targets. Some people lament that irrigation across both India's and Pakistan's Punjab would be better off if operated as per design. Yet, we are in a rapidly changing and uncertain world, with climate change adding to that uncertainty. A more nimble, adaptive management strategy would surely serve irrigation well.

Water bureaucracies, especially irrigation bureaucracies, are notoriously slow to make the change. Many organizations such as the US Bureau of Reclamation, the Mahaweli Authority, or the Water and Power Development Authority of Pakistan were set up primarily as water development organizations—to exploit water to provide access and promote economic development. The International Irrigation Management Institute (IIMI) was set up to help management of newly built but faltering irrigation systems across Asia. Now it is the International Water Management Institute (IWMI), with the name change from 'irrigation' to 'water ' signifying both recognition that irrigation is a key component in a broader field of competing issues, and that water for agriculture goes beyond managing the operation and maintenance of irrigation. As with most institutions, there is a lot of inertia and resistance to change, yet there is perhaps nothing more urgent than this transition. This is well recognized in the many calls for water governance reform.

Global Paradigms—A Help or Hindrance?

The rise of the concept of integrated water resource management (IWRM) in the 1990s (Conca, 2006) corresponded to the fact that many basins in the world had indeed made the transition from open to closed or closing, and new sets of issues arose. The global discourse shifted very much to 'closed basin' topics of reallocation, valuing water, and regulating groundwater, in essence issues of water management, as opposed to water development. A whole new set of tenets for good water management have been developed and discussed and practiced now for several years (Rogers & Hall, 2003). But change is not happening as fast as hoped in spite of this change in paradigm; why not? Going along with the general trend from open to closed basins, and following the international discourse which has started to recognize the importance of management and environment, international water treaties have been increasingly oriented toward management issues in river basins, including environmental conservation and water allocation (Lautze & Giordano, 2007; Giordano & Lautze, 2009). Through the 1960s and into the 1970s, however, many were concerned with the development of hydraulic infrastructure as a means to catalyze much needed economic growth in the continent. While this transition from development to management appears to have occurred throughout much of the world in response to an on-the-ground transition to closed basins, an analogous on-the-ground transition did not take place in most of Africa.

Across much of sub-Saharan Africa, there is a condition of economic scarcity, where water is available, but access is difficult. Freshwater withdrawals are on the order of 70 m^3/capita, while in South Asia they are 700 m^3/capita, and in industrialized countries over 1000 m^3/capita (WCD, 2000; White, 2005). There is a legitimate water development agenda that seems to be well recognized by African leaders. In spite of this apparent need, transboundary treaties in sub-Saharan Africa have followed much the same trend as the world (Figure 1) when it should have been expected that development concerns would have gained more attention. On the other hand we have to recognize that there is a huge international pressure from donors, research agencies, and institutions promoting global paradigms for the management agenda. This pressure is indeed hard to resist by countries in need of development funds. One has to ask whether good intentions are misplaced because they are out of context.

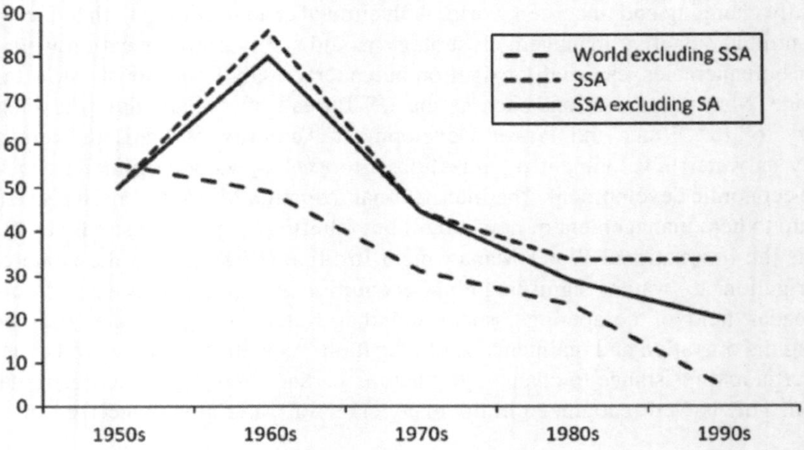

Figure 1. Transboundary law globally is transitioning from concerns of water development to water management incorporating environmental concerns. Africa follows the same trend in spite of water development remaining a key concern. *Source:* Lautze & Giordano, 2007.

Mukherjee (2005) points out that perceptions of groundwater drawn from closed basins impact policies of groundwater in open basins. In the Eastern Indian state of West Bengal the government has applied particularly strict regulations on the use of groundwater in the region. This situation is odd because there is no shortage of groundwater in the area. Mukherjee argues that the perception of groundwater overdraft carries over from water-scarce regions like the Punjab to the water-rich region of the Eastern India. Furthermore, the policies and regulations implemented are designed for a much more formal system with which farmers have little experience. The result is that there is excessive groundwater regulation, and not enough groundwater use, which ultimately hampers the economic development of the region. Tushaar Shah in *Taming the Anarchy* (Shah, 2009) provides several more examples where international prescriptions are questionable. Sri Lanka and Tanzania, following the tenet of treating water as an economic good, introduced a flat tariff. With the intention of updating water governance and formalizing water institutions, dozens of countries across sub-Saharan Africa have declared water state property, and introduced water rights. Participatory irrigation management has been tried now for 20 years. Unfortunately, there is little to show from these efforts.

Formal and Informal Water Economies

A marked feature, especially of agricultural water use in many countries, is the degree to which it falls outside the main scope of water management agencies. Examples abound. Farmers the world over are turning to pumps to irrigate their fields rather than rely on irrigation. The reasons are many—from being able to obtain a reliable supply of water when it is wanted, to not having to rely on an irrigation bureaucracy to deliver the same service. Farmers are willing to pay more in terms of energy to get that water. Pumping from groundwater serves over half of India's irrigated area. Pump sales are booming across Asia (Figure 2).

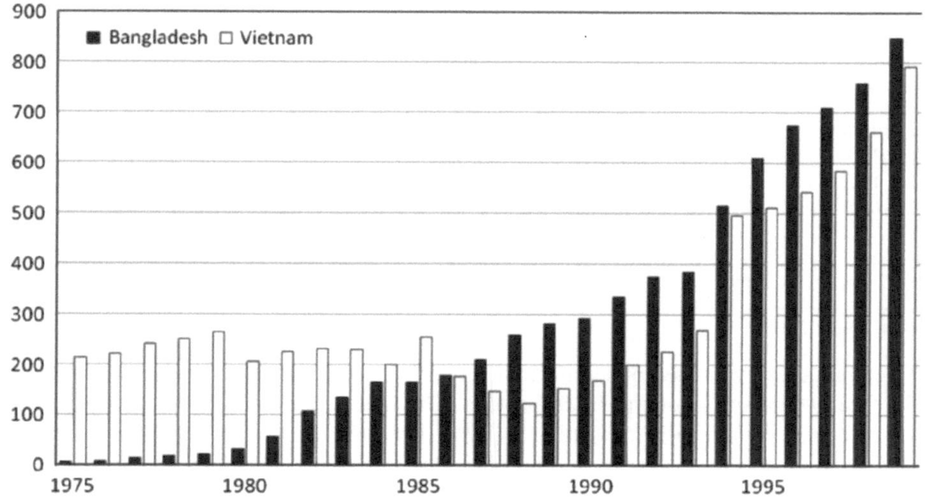

Figure 2. Pumps for irrigation, most often installed and managed outside the formal economy, are becoming increasingly important for agriculture in Asia and across the globe. *Source:* Molden *et al.*, 2007b; graphic reprinted courtesy of Earthscan—www.earthscan.com.

For many groundwater users operating informally, the boom may not last as groundwater tables decline below limits where it becomes uneconomical to pump. In part because they are outside of the formal water economy, they are harder to regulate.

In Ghana, a country with just 5500 ha developed and used under formal irrigation, it was found that around of the city of Kumasi about 12,000 ha of area was irrigated by informal irrigators, unrecognized by government and falling outside of official statistics. In fact, these farmers fell outside the law and were subject to fines where polluted water was used. Later, a new irrigation policy was drafted which gives the informal irrigation sector full recognition (Drechsel *et al.*, 2006), and calls for more progressive health risk reduction options which put farmers less on the spot when they have no alternative to the use of polluted water. This new policy is in line with the 2006 edition of the World Health Organization (WHO) guidelines for safe wastewater use in agriculture (WHO, 2006).

Looking across the globe it was found that wastewater use for agriculture, both treated and untreated, is prevalent in many cities, but a situation that is routinely missed by policy makers (Raschid-Sally & Jayakody, 2008) because of its informal nature. Studying 53 cities, it was found that 1.1 million farmers cultivate 0.4 million hectares of land, and provide an important source of vegetables to cities. On the negative side, these farmers do not get the regular support offered through government agencies. On the positive side, people are experimenting and learning how to cope with water scarcity on their own, offering innovative solutions that may not otherwise have arisen. This presents an interesting example of an informal management scheme dealing with a "closed" basin better.

Table 2 contrasts features of informal and formal water economies. In reality, a mix occurs where part is formal and part informal. For example, in the Western United States, the formal sector would dominate, with some informal arrangements. In contrast, India would have a large informal sector, but with formal bureaucracies. This ratio can shift, as it did in India with many irrigators choosing groundwater over formal large scale irrigation.

Table 2. Characteristics of formal and informal water economies

Informal	Formal
Self-supply predominates	Service providers dominate
Vast numbers of tiny, primary water diverters from nature	Very few, but large primary diverters of water from rivers, lakes
Water institutions are local, fragmented, informal	Water institutions are few, formal, legal bodies
Intermediation in water services are low or absent	Very high degree of intermediation in water provision
Even if water is scarce its free	Even if water is plentiful, it costs money

Source: Shah, 2009.

An advantage to working in the informal economy is that water users do not have to rely on ineffective government mechanisms to access water. Farmers basically take water issues into their own hands, harvesting water or installing pumps to get water when it is needed. Also, informal systems are much faster to react than formal institutions because they are not linked to bureaucracy. In addition to this, prices do not constrain physical access as the water resource is free. A major disadvantage of working within the informal sector is that there are few mechanisms to handle conflict between individuals, or to regulate overexploitation of resources. Furthermore, informal systems can lead to an uneven distribution of water resources because those with the money to buy good pumps can get the most water. This often works to accentuate a rich–poor gap in places with informal water management schemes. Together with the framework of open and closed basins, the framework of formal and informal water economies provides important context for examining a river basin.

Managing Changing Societal Objectives

A major governance challenge of the 21st century will be to deal with the competition for water between agriculture and cities, especially in closed basins. As city water demands grow, irrigated areas are targeted as a possible supply of water (Molle & Berkoff, 2006). The logic is that if agriculture could produce the same amount of food with less supply, more water could be made available for cities. While this water reallocation from presumably wasteful agriculture to cities may be a broad objective for an increasingly urbanized society, it is not likely to be an objective of a farmer whose life is easier with a larger water supply, or who does not want to pay more for a diminishing water supply.

Possible solutions are to manage the demand of farmers and city dwellers through pricing or market-based mechanisms. The logic is that higher prices would provide a stimulus for farmers to conserve water. Urban households and industries would be willing to pay more for water. Water rights could be bought and sold at market prices to encourage water to move to the highest value use. Alternatively, city uses could somehow compensate producers by buying saved water from agriculture. For example, the city of Los Angeles provided funds to the Imperial Valley to transfer conserved water to the city, but maintains sufficient water to keep agricultural land in production (CGER, 1992).

This rural-urban water reallocation situation is very much a reality across China, and the Zhang He area provides some clues on how to deal with this in a different way (Barker *et al.*, 2004). From the 1960s through the 1980s, agriculture dominated as food security was a major concern. The Zhang He Reservoir in Hubei was constructed to irrigate two rice crops per year and to provide hydropower and flood protection. As the economy started to liberalize in the 1980s, there was more urban growth, and agriculture itself made major changes. Over time a remarkable pattern emerged with non-agricultural withdrawals from the reservoirs exceeding agricultural withdrawals. Food security remains a major concern in China, and competition for water between agriculture and cities is a reality. The real success was that the transfer was made at no expense to agricultural production which has not markedly declined (Figure 3).

How did this happen? While Chinese farmers pay for water, it was not the market that provided the real incentive. The Zhang He irrigation system serves thousands of small-holder irrigators through a hierarchical web of canal systems (Loeve *et al.*, 2004). While formal canal operations deliver water from the main reservoir to main canals and to communities and farmers, within communities farmers negotiate water distribution, often in inventive ways, and not necessarily the same. Farmers through their community leaders negotiate with water managers about the distribution of water. Water measurement takes place and there is a charge for water, but is not connected to the volume of water at the farm level. It would be very complicated for operators to consider the demands of thousands of individual farmers. The solution found by water managers at Zhang He relies little on market mechanisms, nor does it rely on high levels of participation. Rather, operators over time have allocated an increasing amount to cities, and less to farmers—a somewhat top down approach. There are incentives. Cities and hydropower both pay a higher fee for water than do agricultural producers. Reduced supplies to farmers provide an incentive for them to respond. Part of the response was the construction of thousands of small reservoirs within the irrigated area to capture runoff generated within the command area, plus to capture return flows from rice cultivation (Roost *et al.*, 2008). In addition, research demonstrated that yields would not suffer if rice fields were not left flooded, and alternating wet and dry irrigation could be employed (Li & Barker, 2004). With the extension of knowledge about alternating wet and dry irrigation of rice, farmers had a technology to help them cope. A remarkable trend has emerged. Crop production has

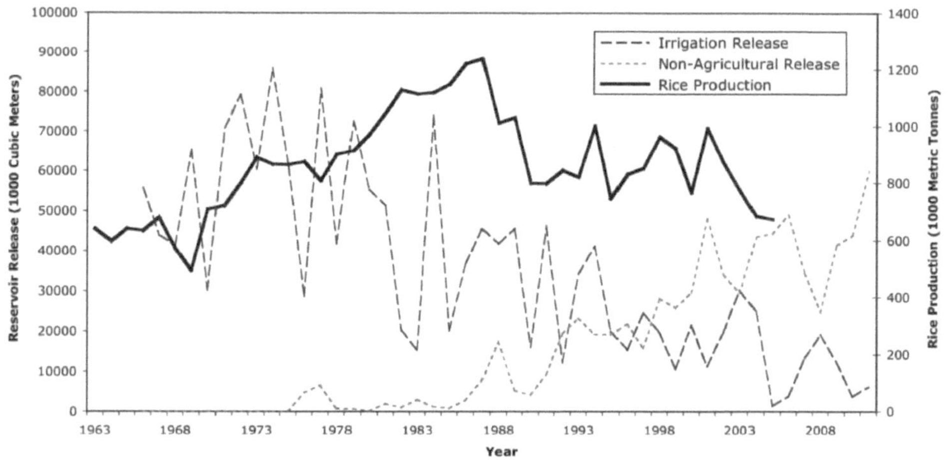

Figure 3. Water Allocation and Rice Production, Zhang He Reservoir, Hubei, China. *Source:* Developed from data in Molden *et al.*, 2007a

remained steady in spite of less water being delivered from the main reservoir to rice cultivators. Productivity of rice, as in most of China, has increased. Water productivity gains in the Zhang He irrigation system have skyrocketed because of increased agricultural productivity, with less reservoir releases.

It is important to note that water productivity gains were not the driving factor to releasing water, rather it was the main response to diminished supplies of water from the main reservoir. Water pricing plays a role for cost recovery, but it was not a major stimulus to save water. The major lever was the declining main system water allocation. Water managers had found a simple lever, low in transaction costs, to stimulate reallocation, and at the same time minimize transaction costs across the system. Farmers were further removed from the formal irrigation system, to more informal arrangements for self-supply with the small reservoirs.

Managing Transitions: A New Conceptual Framework

The examples above have brought out two important areas of consideration (Figure 4). The first is whether the basin of interest is open or closed. The second is the degree of informal or formal institutions to manage water. These two considerations provide important contextual dimensions to aid in understanding the status of how people use water, and suggest why some solutions may be appropriate while others may not. The framework helps conceptualize the various types of transitions that water managers face, including the transition from open to closed basins, changing societal demands, and changing drivers from outside the water sector.

The matrix above depicts graphically in one figure the open–closed and informal–formal transitions. In the upper left hand quadrant are areas where water is still abundant, and many water users are acting relatively independently of the formal sector. The Ganges, an open river basin, has a mix between formal and informal users, with many informal users pumping groundwater, or building impoundments to harvest drainage water or rainwater. The Indus, a closed basin with large formal irrigation, has symptoms of water scarcity with groundwater decline, and competition for water. However, with its heavy

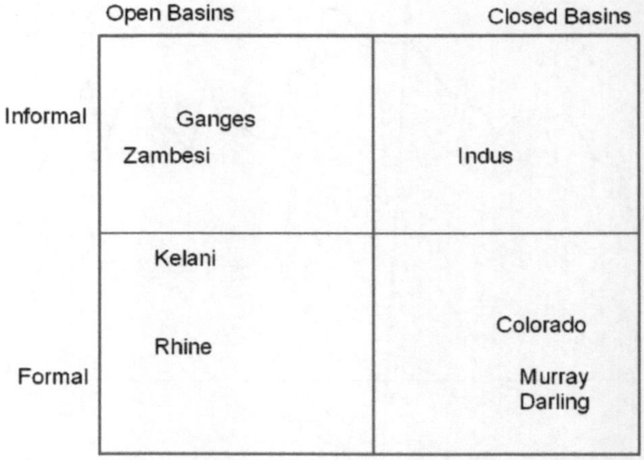

Figure 4. Conceptual framework for managing transitions.

groundwater use, it still has a large informal sector. In fact, it could be argued that the formal water use is losing ground to informal water use in both the Indus and the Ganges.

The lower left quadrant has areas of intensive water use, yet water is relatively abundant relative to demand. Urban and industrial supplies dominate over agricultural water use. There are institutions for planning and regulation of water use. In these areas where water is licensed, it is relatively easier to put in place economic instruments for demand management. The Kelani River, which provides the water supply to Colombo, Sri Lanka, is one where the formal water supply sector dominates over the informal with municipal, hydropower and industry uses dominating over agriculture.

The Murray Darling and Colorado rivers are closed river basins with a high degree of regulation. Water rights exist, water is metered, and economic instruments play an important role in water allocation. Water in these areas is reallocated out of agriculture to cities and environmental programmes using a combination of regulatory approaches and economic instruments. These areas are often cited as models for water management. While these basins indeed provide important examples, they fall into specific quadrants in the matrix elaborated above. In the context of this broader analytical framework, blindly transferring practices in these basins to basins falling into other quadrants seems inappropriate.

In particular, one question is whether areas with a high degree of informal water users should shift to the formal sectors. Advantages of formal water economies are that water can be better regulated, and perhaps provide remedies for competing uses and environmental degradation such as declining groundwater levels. Bringing in more formal users has implications for water governance to take into consideration the voices and needs of many in the informal sectors. It also raises some sticky issues such as who gets rights to water when previously there was no formal water rights system.

A second question is whether basins can jump from informal/open to formal/closed with water development. River basins in Africa are likely to undergo significant changes pushing them from open to closed. However, whether they can leapfrog to situations where best practice IWRM solutions can be implemented remains an open question.

Second Best Solutions

Societies will confront a variety of water problems with water scarcity. While 'best practice' solutions developed elsewhere in the world may provide clues as to how local problems can be solved, solutions must be tailored to local situations. Too often, however, 'best practices' derived from the formal/closed situation do not fit local situations. The framework presented above presents important contextual dimensions, as well as a view of transitions, that can provide more guidance for developing solutions.

India provides an example of how a closed basin with a large informal water economy has devised an alternative solution. In Gujarat, as across much of western and southern India, groundwater overdraft is a threat to the long term sustainability of water resources (Shah *et al.*, 2008). One reason for heavy groundwater use is heavy subsidies on electrical supplies. In fact, the water problem extended to the energy sector, where there were many blowouts of capacitors and brownouts because the water sector was putting so much pressure on energy. A logical and frequently invoked solution includes the formulation of provision of water licenses, metering flows, and water pricing as part of a programme to bring groundwater use under control. Such a programme has not found a place because it would in essence mean establishment of a complicated programme of bringing thousands of groundwater users from the informal to the formal sector. Other drawbacks would be the cost and complexity of metering and measuring water for a dense population of small-holder farmers, and the chance of opening the door to rent-seeking behaviour. Running such a system would involve fairly high transaction costs. The farm lobby strongly politically opposes this method, and several elections have swung on this issue.

A second alternative, the Jyotigram programme, was devised to separate domestic electrical supplies from electricity supplies for water pumps, and to ration electrical supplies to farmers. The advantage to farmers was that it greatly increased the reliability of their water supply and provided better quality power. Rather than in the informal system where they had to use phase splitters much of the time to power their pumps—thus decreasing the quality of the power for all—they were given eight hours of three-phase power each day. The greater advantages were for the villages, where domestic users were given 24-hour three-phase power. This greatly increased the productivity of rural housewives, and improved conditions in schools, hospitals, and all non-farm industries. Furthermore, it provided an easy way to manage groundwater use without the complexity usually associated with a full IWRM scheme (Shah & Van Koppen, 2006).

Participatory irrigation management, market-based solutions, water licensing and pricing, and environmental flows are all important parts of a global toolbox. These have had success in some contexts, but have failed in other situations because of such contexts as described above. Often, the solution that initially seems best able to manage the problem is unfeasible for any number of reasons. 'Second best' solutions are an alternative. The Jyotigram example showed how a policy change in the energy sector could be used to influence the informal water sector. Similarly, the Zhang He example used a relatively simple water allocation approach at the main canal, instead of trying to work with all farmers, in order to indirectly influence the irrigation behaviour of thousands of farmers. In addition, they attempted to support farmers with research and extensions, as well as subsidies for small reservoirs. Moreover, the application of a tried-and-tested solution in one context may prove to be detrimental in solving local problems in another.

Strictly following international best practice and IWRM principles would have resulted in a solution that was not effective for the Gujarat and Zhang He examples. Had a governance process not been initiated in Gujarat, IWRM principles would probably have been pushed by water institutions, and there would have been a political stalemate. However, a governance process demonstrated an alternative and more feasible solution for Gujarat; in a sense, good governance trumped a preordained prescription for good management. This example demonstrates that sometimes good management is not enough, particularly when the good management practice is not specific to a particular location, and that good governance is necessary even in cases where management is excellent. However, this lesson is not all that it seems, especially when compared to the lesson learned in the Zhang He reservoir case in China.

The Zhang He example, by contrast, showed that top-down approaches that are well-planned can be effective. Framed in terms of management and governance, the Zhang He example demonstrates that sometimes water management needs fewer governance processes to function effectively. Had extensive consultations occurred, the outcome may not have been as effective. Rather than taking the Jyotigram programme as an argument for good water governance, perhaps one should take a combined lesson from Jyotigram and Zhang He—the 'best' solution is not necessarily the right one, or, what we *think* is the right solution does not always work *best*.

The framework of Open/Closed and Formal/Informal basins provides an important general perspective on many of the problems that certain basins are likely to experience. While emerging challenges may look similar across the globe, we should step back and realize that their solutions do not need to be. This is particularly true in the context of the increased variability that climate change and population growth will bring to the water sector. While this variability will demand both better governance and better management, above all it will demand an eye for innovative solutions that are tailored to particular situations.

Conclusions

Based on examples and experiences, we have developed a framework to provide an important context for solving water resource problems. The framework recognizes that transitions and change are the norm for water resources management. The framework also recognizes that there is an important water abundance and scarcity dimension, as well as an institutional dimension—the degree to which informal management exists, caution is warranted when applying norms and 'best practices' for water governance from one context to another. In particular, it is important to recognize that 'accepted' norms are often derived from richer country contexts and may not be immediately transferrable. One major reason is the assumption that the formal sector can readily manage water in areas where there is a large informal sector. Developing countries should take advantage of the positive aspects of developed country experience, and look for opportunities to leapfrog. However, to avoid negative consequences, they should watch for limits to leapfrogging by avoiding global policy prescriptions and 'best practices' not built on the experience of local environments.

Caution need not stymie action. There is a case for proactively adapting governance and management arrangements as basins shift from open to closed and as governance shifts from informal to formal. In some cases, it may be appropriate to encourage the shift from the informal to formal. However, it may often be more appropriate to look for advantages in working within the informal water sector, or in understanding how informal and formal may work simultaneously at different scales.

Allow for, and often seek out, second best solutions. The 'ideal' is not likely to be implementable in practice, and even if it is, it may not have the desired result when considered within the overall social, economic and political context. Second best water governance solutions may go against apparent norms, but can push us towards desired results. Use indirect instruments to work towards public policy goals in the informal sectors of the water economy, as illustrated by the Jyotigram example from India. Look for solutions that lower transaction costs, and that that align incentives between various interest groups, such as in the example from Zhang He, China. Be careful of norms developed in closed basins and wealthy countries, as shown in the example of African transboundary waters. Applied broadly, the concept of searching for specific local solutions that do not adhere to the dominant global paradigms of water management can provide unexpected and above all, useful options.

References

Barker, R., Tuong, T. P., Li, Y., Castillo, E. G. & Bouman, B. A. M. (2004) Growing more rice with less water: research findings from a study in China, *Paddy and Water Environment*, 2(185), p. 185.

CGER (Commission on Geosciences, Environment and Resources) (1992) California's Imperial Valley: a "winwin" transfer?, in: *Water Transfers in the West: Efficiency, Equity, and the Environment* (Washington, DC: The National Academy of Sciences).

Conca, K. (2006) *Governing Water: Contentious Transnational Politics and Global Institutions Building* (Cambridge, MA: The MIT Press).

Drechsel, P., Graefe, S., Sonou, M. & Cofie, O. O. (2006) *Informal Irrigation in Urban West Africa: An Overview*, Research Report 102, Colombo, Sri Lanka: IWMI. Available at http://www.iwmi.cgiar.org/Publications/IWMI_Research_Reports/PDF/pub102/RR102.pdf.

Falkenmark, M. & Molden, D. (2008) Wake up to realities of river basin closure, *International Journal of Water Resources Development*, 24(2), pp. 201–215.

Giordano, M. & Lautze, J. (2009) Managing waters in extreme environments: the role of international actors in Africa, in: C. Lichpin, D. Sandler & E. Cushman (Eds) *Managing Waters in Extreme Environments* (Dordrecht: Springer).

Giordano, M., Wolf, A. & Giordano, M. (2008) Institutions for transboundary basins, in: C. Sadoff, T. Greiber, M. Smith & G. Bergkamp (Eds) *Share: Managing Water Across Boundaries*, pp. 65–78 (Gland, Switzerland: International Union for Conservation of Nature).

Giordano, M. & Villholth, K. (2007) The agricultural groundwater revolution: setting the stage, in: M. Giordano & K. Villholth (Eds) *The Agricultural Groundwater Revolution: Opportunities and Threats to Development*, pp. 1–4 (Wallingford, UK: CABI).

Hooper, B. (2006) *Key Performance Indicators of River Basin Organizations* (Alexandria, VA: US Army Corps of Engineers).

Iza, A. & Stein, R. (Eds) (2009) *Rule: Reforming Water Governance* (Gland, Switzerland: IUCN).

Keller, J., Keller, A. & Davids, G. (1998) River basin development phases and implications of closure, *Journal of Applied Irrigation Science*, 33(2), pp. 145–163.

Lautze, J. & Giordano, M. (2007) Demanding supply management and supplying demand management: transboundary waters in Sub-Saharan Africa, *Journal of Environment and Development*, 16(3), pp. 290–306.

Li, Y. & Barker, R. (2004) Increasing water productivity for paddy irrigation in China, *Paddy and Water Environment*, 2(4), pp. 187–193.

Loeve, R., Hong, L., Dong, B., Mao, G., Chen, C. D., Dawe, D. & Barker, R. (2004) Long-term trends in intersectoral water allocation and crop water productivity in Zhanghe and Kaifeng, China, *Paddy and Water Environment*, 2(4), pp. 237–245.

Molden, D., Sakthivadivel, R., Samad, M. & Burton, M. (2005) Phases of river basin development: the need for adaptive institutions, in: M. Svendsen (Ed.) *Irrigation and River Basin Management: Options for Governance and Institutions*, pp. 19–20 Wallingford, UK: CABI and Colombo, Sri Lanka: International Water Management Institute (IWMI).

Molden, D., Bin, D., Loeve, R., Barker, R. & Tuong, T. P. (2007a) Agricultural water productivity and savings: policy lessons from two diverse sites in China, *Water Policy*, 9(Supplement 1), pp. 29–44.

Molden, D., Frenken, K., Barker, R., de Fraiture, C., Mati, B., Svendsen, M., Sadoff, C., Finlayson, M., Atapattu, S., Giordano, M., Inocencio, A., Lannerstad, M., Manning, N., Molle, F., Smedema, B. & Vallee, D. (2007b) Trends in water and agricultural development, in: D. Molden (Ed.) *Water for Food, Water for Life: A Comprehensive Assessment of Water Management in Agriculture*, pp. 57–89 (London, UK: Earthscan and Colombo, Sri Lanka: IWMI).

Molle, F. (2003) *Development Trajectories of River Basins: A Conceptual Framework*, IWMI Research Report 72 (Colombo, Sri Lanka: International Water Management Institute (IWMI)).

Molle, F. & Berkoff, J. (2006) *Cities Versus Agriculture: Revisiting Intersectoral Water Transfers, Potential Gains and Conflicts*, Comprehensive Assessment Research Report 10, Colombo, Sri Lanka: International Water Management Institute. Available at http://www.iwmi.cgiar.org/Assessment/files_new/publications/CA%20Research% 20Reports/CARR10.pdf

Molle, F., Mamanpoush, A. & Miranzadeh, M. (2004) *Robbing Yadullah's Water to Irrigate Saeid's Garden: Hydrology and Water Rights in a Village of Central Iran*, Research Report 80, Colombo, Sri Lanka: International Water Management Institute.

Mukherji, A. (2005) Political ecology of groundwater: the contrasting case of water-abundant West Bengal and water-scarce Gujarat, India, *Hydrogeology Journal*, 14(3), pp. 392–406.

Raschid-Sally, L. & Jayakody, P. (2008) *Drivers and Characteristics of Wastewater Agriculturein Developing Countries: Results from a Global Assessment*, IWMI Research Report 127, Colombo, Sri Lanka: International Water Management Institute.

Rauch, T. (2009) *The New Rurality: Its Implications for a New, Pro-Poor Agricultural Water Strategy* (Rome: IFAD).

Rogers, P. & Hall, A.W. (2003) *Effective Water Governance*, TEC Background Paper 7, Global Water Partnership.

Roost, N., Cai, X. L., Turral, H., Molden, D. & Cui, Y. L. (2008) Adapting to intersectoral transfers in the Zhanghe irrigation system, China—Part II: impacts of in-system storage on water balance and productivity, *Agricultural Water Management*, 95(6), pp. 685–697.

Seckler, D. (1996) *The New Era Of Water Resources Management: From "Dry" To "Wet" Water Savings*, Research Report 1, Colombo, Sri Lanka: International Irrigation Management Institute (IIMI).

Shah, T. & Van Koppen, B. (2006) *Is India Ripe for Integrated Water Resources Management (IWRM)? Fitting Water Policy to National Development Context*. Colombo, Sri Lanka: International Water Management Institute.

Shah, T. (2009) *Taming the Anarchy: Groundwater Governance in South Asia* (Washington, DC: Resources for the Future).

Shah, T., Bhatt, S., Shah, R. K. & Talati, J. (2008) Groundwater governance through electricity supply management: assessing an innovative intervention in Gujarat, western India, *Agricultural Water Management*, 95(11), pp. 1233–1242.

Shah, T., Burke, J., Villholth, K., Angelica, M., Custodio, E., Daibes, F., Hoogesteger, J., Giordano, M., Girman, J., van der Gun, J., Kendy, E., Kijne, J., Llamas, R., Masiyandima, M., Margat, J., Marin, L., Peck, J., Rozelle, S., Sharma, B., Vincent, L. & Wang, J. (2007) Groundwater: a global assessment of scale and significance, in: D. Molden (Ed.) *Water for Food, Water for Life: A Comprehensive Assessment of Water Management in Agriculture*, pp. 395–423 (London, UK: Earthscan).

Tropp, H. (2007) Water governance: trends and needs for new capacity development, *Water Policy*, 9(S2), pp. 19–30.

White, W. R. (2005) *World Water Storage in Man-Made Reservoirs* (Marlow, U.K. Foundation for Water Research).

World Commission on Dams (WCD) (2000) *Dams and Development: A New Framework for Decision-Making* (London: Earthscan Publications).

Seckler, D. (1996) *The New Era of Water Resources Management: From "Dry" to "Wet" Water Savings* (Colombo, Sri Lanka: International Irrigation Management Institute).

World Health Organization (2006) *Guidelines for the Safe Use of Wastewater, Excreta and Greywater*, Volume 2 (Geneva: WHO).

Irrigation Modernization in Spain: Effects on Water Quantity and Quality—A Conceptual Approach

S. LECINA*, D. ISIDORO**, E. PLAYÁN*** & R. ARAGÜÉS**

*Ager ingenieros, ingeniería rural y civil S.L. c/Mayor 53, 2°B 50001 Zaragoza, Spain; **Soil and Irrigation Department, Agrifood Research and Technology Centre of Aragon, Government of Aragon, Zaragoza, Spain; ***Department of Soil and Water, Aula Dei Experimental Station, Zaragoza, Spain

ABSTRACT *This article analyses the effects of irrigation modernization processes on water quantity and quality, taking the Ebro river basin (NE Spain) as a case study. The objective is to contribute to needed in-depth analysis and discussion regarding the optimization of water use in agriculture. A conceptual approach based on water accounting concepts has been applied. Results show that irrigation modernization linked to an increase in land productivity involves additional water depletion if the location of the irrigated areas and the quality of the irrigation return flows allow their re-use. Additionally, modernization reduces the volume of return flows and pollutant loads and increases the quality of the receiving water bodies. The modernization of water management will be required to maximize economic, social and environmental returns from the investment in new irrigation infrastructure.*

Introduction

Water Scarcity Management in Spain

Spain has a Mediterranean climate over most of its territory. Relevant infrastructure has been built throughout its history for storing and delivering water. In addition, a pioneering legal framework was developed for regulating water use. The concept of river basin unity for water resources management was introduced in the early 20th century. Water users associations were introduced in the first Water Law that was drawn up in the 19th century (Ministerio de Fomento, 1879).

Spain's economic growth in recent decades has substantially increased water demand. However, water availability has hardly increased because of the lack of significant increments in water storage capacity. These facts have strengthened competition for water resources, and cyclical droughts have brought social conflicts between users and regions within Spain (MARM, 2006).

The Spanish government has introduced reforms to manage water demand. Public water rights banks, environmental taxes, and subsidies for irrigation modernization are just some examples. In addition, new water plans are being implemented through a participative and integrated process following the guidelines of the European Water Framework Directive (European Union, 2000).

One of the most ambitious plans is irrigation modernization. Spain has approximately 3.5 Mha of irrigated land. Although this surface represents only 13% of total agricultural land, it generates about 50% of the agricultural gross domestic product (Forteza del Rey, 2002). Before the onset of these plans, surface irrigation amounted to 59% of the irrigated area, and 71% of the area had infrastructure more than 25 years old (MARM, 2002). About 72% of the irrigated area is located in land-locked provinces (INE, 1999). Surface irrigation is predominant in these provinces, where field crops occupy 74% of the surface irrigated area (MARM, 2007).

The two National Irrigation Modernization Plans (*Plan Nacional de Regadíos* and *Plan de Choque de Modernización de Regadíos*) were designed with two main objectives: 1) to increase the competitiveness of the irrigation sector in order to face the progressive liberalization of agricultural markets and the reduction of subsidies; and 2) to save 3,000 Mm^3 of water per year to alleviate the consequences of cyclical droughts in other sectors. This water saving represents 15% of the yearly average national agricultural water use. These plans aim to invest €7,400 million during this decade to improve nearly 2 Mha of irrigation infrastructure of (MARM, 2002; MARM, 2006).

Analysis of Agricultural Water Uses in Spain: Irrigation Efficiencies

The concept of efficiency has been widely used to evaluate agricultural water use in Spain. Efficiencies have traditionally been applied to irrigation system design and scheduling, and for computing maximum allowable withdrawals for irrigation. The water saving prospect of 3,000 Mm^3 is based on the reduction of water use due to expected improvements in on-farm irrigation efficiency.

The efficiency approach divides the volume used to produce a product or service into two fractions. One is the water beneficially used for obtaining the product or service. The other is the water lost in the process. The relationship between each one of these partial volumes and the total volume determines the respective percentages of efficiency and water losses (Burt *et al.*, 1997).

This methodology to evaluate water use is straightforward and intuitive. However, the efficiency approach does not take into account issues such as water re-use, the distinction between total water use and water consumption, and the influence of location of use within the basin and water quality. Irrigated agriculture is a non-point source of pollutants that causes quality degradation in the water bodies receiving the irrigation return flows, decreasing its capability to re-use (Ongley, 1996).

These issues are particularly important for water management in the context of water scarcity in a basin. For these reasons, several authors have pointed out that the efficiency concept is not appropriate for assessing the hydrological impact of irrigation in a basin (Willardson *et al.*, 1994; Seckler, 1996; Perry, 1999, 2007; Jensen, 2007).

The Water Accounting Approach

Although several authors have included the above mentioned hydrological issues in a new formulation of efficiency called "effective efficiency" (Keller *et al.*, 1996; Seckler *et al.*,

2003; Mateos, 2008), water accounting has been proposed as an alternative for hydrological purposes (Molden & Sakthivadivel, 1999; Clemmens *et al.*, 2008; Perry *et al.*, 2009). This methodology applies the law of conservation of mass through water balances.

Balances identify the destination of the water used and distinguish four sinks of irrigation water (Molden & Sakthivadivel, 1999; Perry *et al.*, 2009): (1) beneficial evapotranspiration; (2) non-beneficial evapotranspiration; (3) non-recoverable runoff/percolation; and (4) recoverable runoff/percolation. The first two elements in relation to the total water use constitute the consumed fraction. Evapotranspiration and non-recoverable runoff/percolation make up the fraction of the total water use that is depleted in a basin.

Objectives

The objective of this work is to contribute to in-depth analysis and discussion regarding the optimization of water use in agriculture. The results of this analysis are expected to provide insight on the consequences of irrigation modernization on water availability and quality.

The consequences of this modernization process on water quantity and quality are analyzed through a conceptual approach applied to the Ebro river basin (ERB) as a case study. The application of basic hydrological concepts to irrigation modernization constitutes a secondary objective of this work. The study is based on the authors' 30-year research experience in the Ebro river basin as well as in a number of experiences around the world.

A clear understanding of the farmers' motivation for irrigation modernization is needed to anticipate the consequences of this process. The following two sections describe the irrigated areas of the ERB and explain their socio-economic context. Subsequent sections analyse the impacts of irrigation modernization on water quantity and quality in the basin and policy implications.

The Irrigated Areas of the Ebro River Basin

The Ebro river basin is one of the most important watersheds of the Iberian Peninsula, covering 15% (85,566 km^2) of its territory. Located in northeast Spain (Figure 1), it has an irrigated area of approximately 800,000 ha (24% of total Spanish irrigated land).

About 73% of the irrigated area, including the largest irrigation projects, is located in the central ERB. The climate in this part of the basin is semi-arid, particularly in the area surrounding the Ebro river. Mean annual temperature is 14 °C, and annual precipitation oscillates between 300–450 mm. A dry period typically extends from July to September. The annual reference evapotranspiration (Hargreaves & Samani, 1985), varies from 949 to 1,157 mm. The average wind speed (at 2.0 m height) ranges between 0.6 m/s and 2.6 m/s.

Soils of the river terraces, where the oldest irrigated areas of the ERB are located, generally cover limestone gravel deposits. They are well drained and present favourable characteristics for surface irrigation. The irrigation projects developed in the 20th century extended the irrigated land outside of these areas, covering platforms, slopes and alluvial terraces. Platform soils sit on tertiary materials covered with gravel. These soils are highly productive because of their low slope and adequate drainage, but often present low surface irrigation efficiency due to their low available water holding capacity (AWHC) and high infiltration (Playán *et al.*, 2000). Soils on slopes and alluvial terraces are characterized by high AWHC but poor drainage. Some of them are naturally salt-affected and others are salt-affected because of levelling, seepage, or particularly low internal drainage.

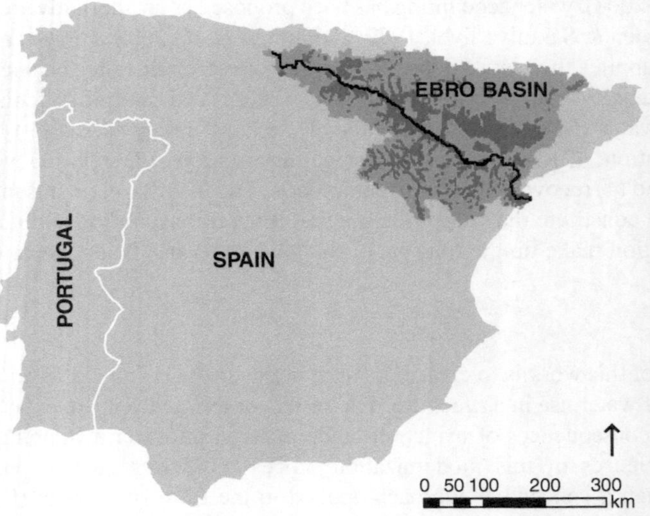

Figure 1. Location of the Ebro river basin in the Iberian Peninsula. *Note*: The darker grey represents the irrigated areas in the basin.

Spots of saline-sodic soils occur in these soils, although sodicity is often associated with irrigation in areas lacking gypsum (Herrero & Snyder, 1997).

Irrigation technology is variable due to the long history of irrigation development in the region. Small irrigated areas were established close to the rivers during the last 2,000 years. The first big canals were built parallel to the Ebro river and the Segre river in the 18th and 19th centuries (about 87,100 ha). The great expansion of the irrigated area took place in the mid-20th century, when massive surface irrigated projects based on large canals and reservoirs built mainly on Pyrenean rivers were promoted by the government (about 327,300 ha). Sprinkler and drip irrigated areas were developed in the last 40 years. Before the establishment of the irrigation modernization plans in 2002, approximately 70% of the irrigated area was irrigated by traditional surface irrigation systems, 19% by sprinkler systems, and 11% by drip irrigation (CHE, 2005).

Figure 2 shows the cropping pattern in the irrigated area of the ERB as an average of the 1996 to 2002 seasons. This cropping pattern was obtained from the government databases used to pay the European Common Agricultural Policy (CAP) subsidies. This period is prior to the beginning of the irrigation modernization process. Field crops occupied 58% of the total area. Fruit trees and vegetables were present in 19% of the area, and olive trees and vineyards were cultivated in 4%. Approximately 17% of the irrigated area was not cropped, mainly due to the archaic structures of the oldest irrigated areas, the low water conveyance capacity of the networks of ditches built in the mid-20th century, the land set aside for CAP subsidies, and the presence of salt-affected soils (Nogués *et al.*, 2000).

Rural Society and Irrigation Modernization in the Ebro River Basin

The agrifood system based mainly on irrigated areas generates about 12% of the gross added value and 14% of the employment in the basin. Agriculture and subsidiary sectors

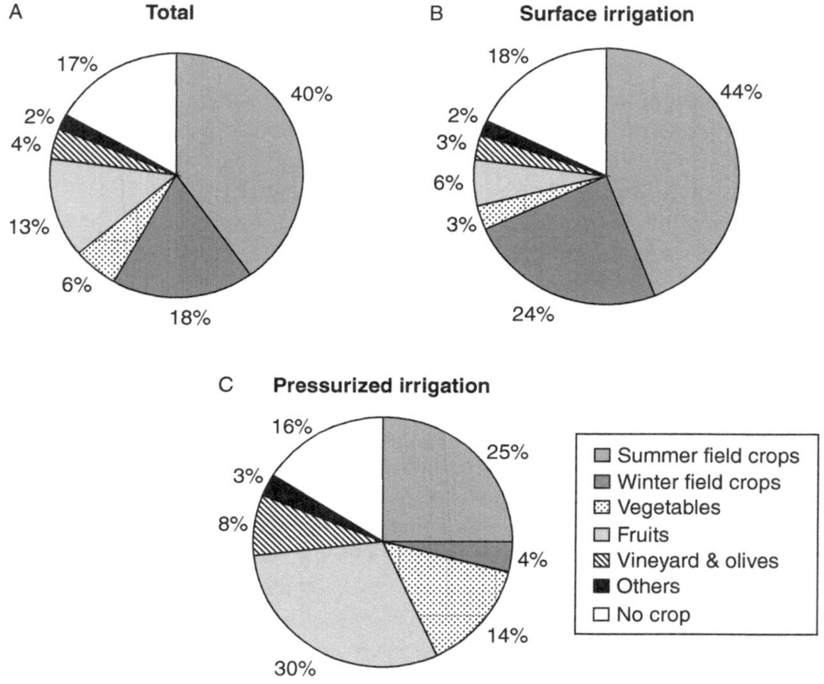

Figure 2. Cropping patterns by irrigation systems in the Ebro river basin (average from 1996 to 2002).

are virtually the only economic activity in many rural regions (CHE, 2005). The CAP has supported agricultural production and its commercialization during recent decades. Agricultural subsidies have been a key factor in farm profits in the ERB, particularly in areas specializing in field crop production. The ongoing liberalization of agricultural trade is bringing sharp decreases in subsidies applied to the crop area, among other measures. This process has resulted in a reduction of subsidies to field crops of about 75% in the last five years.

The liberalization of agricultural trade and the decreased influence of CAP in agricultural markets are also causing an increase in the variability of crop prices. These prices are now much more influenced by multiple and changing factors in time and space, such as world weather, economic and population growth, energy prices, and investment in rural development. These factors have increased the average annual variability of the grain prices at the local agricultural stock exchange, Lonja del Ebro, from 5% to 20% in the last five years. This volatility is expected to continue in the coming years, increasing the uncertainty of the farm net profits (OECD & FAO, 2009).

Spanish economic growth during recent decades has led to a large reduction in the availability and quality of agricultural labour. Young people in rural areas prefer urban jobs because these have higher wages and levels of technology than agricultural jobs. This trend is evident by the low population density characterizing the basin. Overall the average population density is 33 persons per square kilometre, but it is less than 5 persons in 40% of the territory (CHE, 2005).

About 92% of agricultural water comes from surface water supplies, mainly from the Pyrenean rivers, and only 8% from groundwater. Average agricultural water use is about 8,000 m^3/ha (accounted from the water source) (MARM, 2000). The change of land uses in headwater regions of the basin, from crops and pastures to scrubs and forests, due to the depopulation of this territory, has led to a marked reduction in water availability. In the case of the central Spanish Pyrenees mountains these changes have resulted in stream flow reductions of about 30% since the mid 20th century (Beguería *et al.*, 2003). Moreover, changes in precipitation, temperature, and snow accumulation are further contributing to these reductions. If current trends in land cover and climatic conditions continue in the present century, the balance between available resources and water demand could be seriously threatened (López-Moreno *et al.*, 2008).

Irrigated agriculture may contribute to the degradation of water quality because of the salt and agrochemical loadings in irrigation return flows. Although water quality is generally adequate for irrigation in the rivers of the ERB with regard to salinity, nitrate and phosphorus (Isidoro & Aragüés, 2007), local problems have been described in reservoirs, lakes and aquifers (CHE, 2005). Nine zones vulnerable to nitrate pollution from agricultural sources have been declared. Environmental restrictions on agricultural activities are applied in these zones. These restrictions will undoubtedly be stressed in coming years in the basin since a "good environmental status" must be achieved in all water bodies of the basin before 2015 according to the European Water Framework Directive (European Union, 2000).

The productive structure and management of the farms must be competitive, flexible and environmentally friendly to cope with this socio-economic context (Playán & Mateos, 2006). Increasing productivity is essential to achieve farm sustainability. Investments in irrigation should consider these requirements, among others (Turral *et al.*, 2010). Irrigation modernization, including both structural and managerial aspects, stands as an adequate alternative to the current situation of irrigated areas in the central ERB (Albiac *et al.*, 2007; Playán *et al.*, 2007).

These facts stand, therefore, as the main reason why farmers invest in modernizing their irrigation structures. Pressurized irrigation systems provide higher productivities than traditional surface irrigation systems in the ERB. More intensive cropping patterns and higher yields explain this difference.

Before the beginning of the irrigation modernization process in the ERB, 42% of the surface irrigated area was occupied by low value crops (winter field crops like barley and wheat) and fallow plots, in contrast to only 20% in the pressurized irrigated area (Figure 2). Horticultural crops, orchards, vineyards and summer field crops (mostly corn and alfalfa) were present in 77% of the sprinkler- and drip-irrigated land.

Figure 3 illustrates the difference in gross land productivity between irrigation systems by crop groups in the ERB. Gross land productivity was computed as the ratio between the gross value of production and the cropped area (Molden *et al.*, 1998). Average yields and prices were obtained by survey among irrigation district managers and from government statistics during 2003 and 2004 in the central ERB. These data were extrapolated to the whole basin for homogeneity in comparison (MARM, 2004; MARM, 2005). Gross land productivities are about 37% higher for sprinkler and drip irrigated crops due to higher yields. Only vineyards present similar productivities between irrigation systems due to severe yield limitations imposed to achieve quality certified wines.

In addition to these factors, irrigation automation reduces labour requirements. In the old surface irrigation systems in the ERB, one person can irrigate about 50 ha, but more

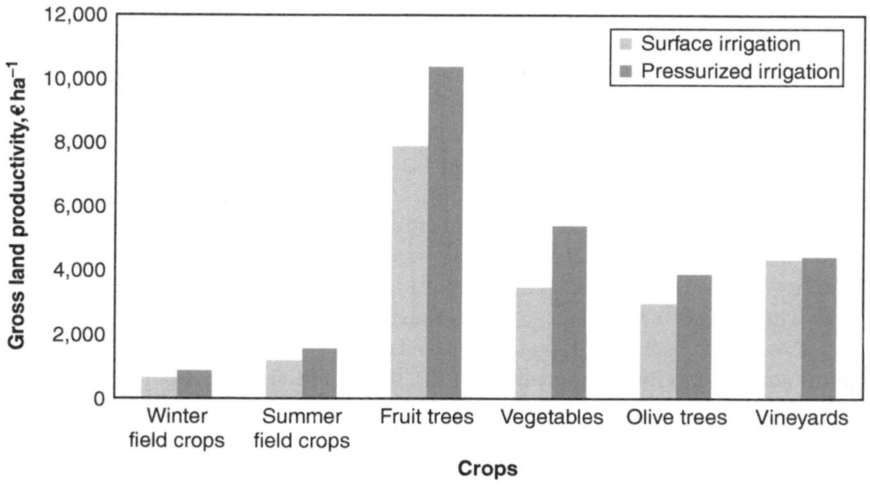

Figure 3. Gross land productivity by irrigation systems and crop groups in the Ebro river basin (average of 2003 and 2004).

than 200 ha can be irrigated by one person using the new pressurized systems. Irrigation modernization leads to increased labour productivity, reduced labour tensions, and improved working conditions for farmers.

On average, a farmer's cost for switching from surface to pressurized irrigation systems (sprinkler or drip) is about €9,000 per hectare, a value similar to the market value of the land. Additionally, energy-related water costs are higher in pressurized irrigation.

Although properly designed and managed surface irrigation systems can perform just as well as pressurized systems (Clemmens & Dedrick, 1994), farmers choose to change the irrigation system in most cases. Irregular topography, small plot size, high labour requirements, and low flexibility are some factors that reduce the productivity of surface irrigation systems in the ERB.

During this decade, approximately 175,000 ha of land in the ERB are expected to have changed surface irrigation systems (MARM, 2002; MARM, 2006). This area is mainly located in the large surface irrigation projects developed in the mid-20th century. In older irrigated areas established before the mechanization of agriculture, the surface irrigation systems are seldom changed because the land holdings tend to be very small, among other reasons. Nevertheless, some of them are improving their networks of irrigation ditches. The government partially supports these efforts in irrigation modernization with subsidies and/or financial aid.

The irrigation modernization process in the ERB will, therefore, result in an increase in crop production and a change in water use practices. These changes will have strong hydrological implications in the basin.

Consequences of Irrigation Modernization on Water Quantity

Beneficial Evapotranspiration

Beneficial evapotranspiration refers to crop evapotranspiration. Proper irrigation scheduling and water application are needed to satisfy crop water requirements.

The old surface irrigation conveyance networks in the ERB were designed at least 50 years ago for low-productivity agriculture based on winter cereals. The low conveyance capacity of the old networks of ditches, coupled with the large irrigation depth generally applied by surface irrigation, often do not allow for adequate irrigation scheduling. Old surface irrigation projects usually present excessively long irrigation intervals. Faci *et al.* (2000) and Lecina *et al.* (2005) describe how farmers tend to apply large irrigation depths because of the uncertainty about when they will irrigate again in the surface irrigated areas of the central ERB. These large irrigation depths do not improve crop water supply and extend the irrigation interval. These authors reported that the intervals are about 10–14 days during the peak months, despite 24-hour daily operation periods.

In soils with low available water-holding capacity, crop water requirements are satisfied only partially due to these long irrigation intervals. If the water available for the crop is not enough to meet its maximum transpiration rate, yields fall below potential levels because of the linear relationship between crop biomass and transpiration (Steduto *et al.*, 2007; Molden *et al.*, 2010).

Isidoro *et al.* (2004), applying water balances in the surface irrigated area of the Violada canal, reported that evapotranspiration is 16% lower than potential. Lecina & Playán (2006a, 2006b) obtained differences of about 15% using combined irrigation-crop models in the Bardenas canal, another surface irrigated area in the central ERB. These differences between real and potential evapotranspiration have been found in other surface irrigated areas in the world (Allen *et al.*, 2005).

In contrast, Tedeschi *et al.* (2001) and Cavero *et al.* (2003) pointed out that evapotranspiration is close to potential in the modern sprinkler-irrigated area of the Monegros II scheme, according to the water balances carried out in this area of the central ERB. New pressurized networks present higher capacity, reliability, and flexibility in water delivery than traditional irrigation ditches. Furthermore, properly designed and managed sprinkler and drip irrigation systems allow farmers to better adjust irrigation depth and frequency to meet the local soil, crop and meteorological features. Zapata *et al.* (2009) reported the capability of these pressurized networks and irrigation systems to achieve a high quality irrigation performance in the new irrigated areas of the central ERB. Therefore, crop water requirements are adequately satisfied and crop evapotranspiration is close to potential in these areas.

The characteristics of the new pressurized networks also allow farmers to grow more productive crops—such as horticultural, orchard or summer field crops—than in traditional surface irrigated areas. These crops generally present higher water requirements (although there are exceptions such as vineyards and olive trees). In some areas (where summer seasons are long enough), it is even possible to grow two or three crops per season.

The differences between irrigation systems in the ERB underscore the point that better water supply means higher crop evapotranspiration and higher crop yields because of the above mentioned biological/crop-water relations. Figure 4 shows average crop evapotranspiration per unit area and gross land productivity by irrigation system in the whole ERB. The cropping pattern presented in Figure 2, differentiated by region and irrigation system, and the gross land productivities shown in Figure 3, were used. In Figure 4, evapotranspiration corresponds to the 50% return probability obtained from Martínez-Cob (2004) for each crop and region in the ERB following the methodology proposed by Allen *et al.* (1998). An average 15% reduction was applied to the potential evapotranspiration in surface irrigated areas regarding the above-mentioned

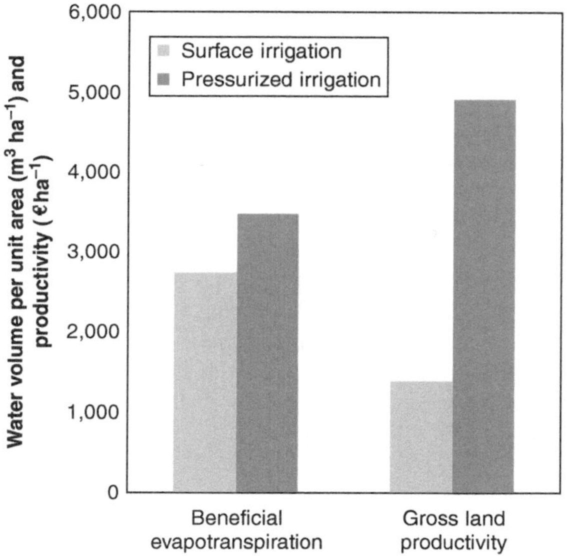

Figure 4. Beneficial evapotranspiration per unit area (50% return probability) and gross land productivity by irrigation systems in the Ebro river basin (cropping pattern referred to the average from 1996 to 2002.

constraints of the low conveyance capacity of the old networks of ditches. The evapotranspiration per unit area in sprinkler areas was 27% higher than in surface irrigated areas, and land productivity was 3.5 times higher. Approximately one half of this difference in evapotranspiration is due to the different cropping patterns.

Water productivity, computed as the ratio between gross land productivity and beneficial evapotranspiration per unit area (Molden *et al.*, 1998; Playán & Mateos, 2006), is €0.508/m^3 in surface irrigated areas, and €1.415/m^3 in pressurized areas. This difference implies a competitive advantage of pressurized systems in drought years, and a more profitable use of the water resources in the basin.

Non-beneficial Evapotranspiration

Non-beneficial evapotranspiration includes evapotranspiration from non-productive plants (like weeds or phreatophytes) and direct evaporation from water bodies. This non-beneficial consumption is common in surface irrigation areas where ditches are present, due to leakages and open water surfaces. The volume of these water flows mainly depends on the maintenance of the structures.

Pressurized networks reduce the volume of water used by non-productive vegetation because they virtually eliminate leakage. Additionally, where this vegetation is present, dikes are not required inside sprinkler- and drip-irrigated plots in contrast to borders and basins in surface irrigation. Pipelines also eliminate direct evaporation. However, pressurized irrigated areas usually have small reservoirs (typically less than 1 Mm^3 in capacity) to increase water storage within them. These reservoirs increase the water surface exposed to evaporation. Nevertheless, Krinner *et al.* (1994) report that the relative volume of this non-beneficial water consumption is generally low in large surface and pressurized irrigation projects of the ERB and other Spanish basins.

In the specific case of sprinkler-irrigated areas, wind drift and evaporation losses (WDEL) can also be considered as non-beneficial water consumption (Burt *et al.*, 1997). Although a small part of these losses offsets crop evapotranspiration due to a modification in plot microclimate (Martínez-Cob *et al.*, 2008), WDEL may notably increase the nonbeneficial component of the balance. Several authors have pointed out that WDEL can vary between 10% and 20% of the total water applied in the sprinkler irrigated areas of the central ERB (Dechmi *et al.*, 2003; Playán *et al.*, 2005). This magnitude depends on the design and management of the irrigation system, and on weather conditions. The central ERB, particularly near the Ebro river, is characterized by strong winds, a key factor effecting WDEL. In the conditions of this area, Zapata *et al.* (2007) report that decreasing WDEL below 10% is strongly limited by the cost of the required structures and the low value of field crops.

Runoff/Percolation

Runoff/percolation comes from irrigated plots and conveyance structures. Field evaluation campaigns carried out in surface irrigated areas of the Bardenas canal obtained on-farm application efficiencies (Burt *et al.*, 1997) of about 40% in areas where soils are characterized by low AWHC and high infiltration rates (Lecina *et al.*, 2005). Where soils present more favourable conditions for surface irrigation, application efficiencies were about 60% (Lecina *et al.*, 2005). Similar values were obtained by Playán *et al.* (2000) and Lecina *et al.* (2000) in other surface irrigated areas of the ERB. Water balance measurements carried out in the Violada canal and the Bardenas canal obtained return flows equivalent to approximately 50% of the water inflows (Isidoro *et al.*, 2004; Causapé *et al.*, 2004a; Causapé *et al.*, 2004b).

In contrast, field evaluation campaigns performed in pressurized irrigated areas of the Cinca canal and the Ebro river showed average on-farm efficiencies around 80% (Dechmi *et al.*, 2003). Tedeschi *et al.* (2001) and Cavero *et al.* (2003) obtained return flows which represented between 5 and 11% of the inflows in water balances carried out in the sprinkler-irrigated area of the Monegros II scheme.

The authors of these studies report that virtually all of the runoff/percolation volume returns to the rivers. The quality of these returns allows other users to reuse them, as will be explained in the next section. Approximately 30% of these flows are re-used within the same surface irrigation projects where they are generated (Causapé *et al.*, 2006; Causapé, 2009). The remaining irrigation return flows can be re-used downstream by other users along the basin. Only 6% of the irrigated area in the ERB is located at a distance lower than 100 km from the Mediterranean Sea, following the course of the Ebro river. Salt lakes or other sinks where water is non-recoverable inside the ERB are small and have little impact on irrigation return flows according to Castañeda & García-Vera (2008) and CHE (2005). These facts result in a low non-recoverable runoff/percolation.

Water Depletion

The reusability of irrigation return flows, and the low volume of non-beneficial evapotranspiration, convert beneficial evapotranspiration into the main component of the depleted fraction, and leave a limited margin for real water savings in the ERB. The change of irrigation system and the intensification of crop production as a consequence of irrigation modernization will result in an increase in water depletion despite the improvement of on-farm efficiencies, particularly in the cases of switching from surface to sprinkler irrigation owing to WDEL. The overall effect on water demand will depend on whether the increase in evapotranspiration is higher or lower than the decrease in runoff/percolation.

Considering the cropping pattern by irrigation system prior to modernization (Figure 2), the planned change of irrigation systems covering 175,000 ha will involve an increase of about 141 Mm^3 (from 2,426 Mm^3 to 2,567 Mm^3 a year) in the beneficial evapotranspiration of the ERB (50% return probability). WDEL should be added to this volume in the cases of switching from surface to sprinkler irrigation to estimate the total increase in water consumption. This increment implies an equivalent reduction of water availability in the basin.

There are many factors influencing the cropping pattern, and therefore water consumption. Winter cereals and fallow plots will not be completely eliminated in modernized areas because rotation of field crops is needed for weed control, and salt-affected areas barely go through modernization. In a fully liberalized market, the international prices for agricultural commodities and energy will dynamically affect the degree of crop intensification (IBRD, 2009). Additionally, future water restrictions and growing market competition could force marginal farms (that is, non-modernized farms) out of production. The foreseen agricultural context could encourage the application of regulated deficit techniques (RDI) in the future.

Consequences of Irrigation Modernization on Water Quality

Water pollution is a key factor in water availability. In land-locked irrigated areas evapoconcentration and weathering are two significant water quality degradation processes induced by irrigation (Aragüés & Tanji, 2003). The impact of irrigation modernization in water quality will depend on the relative importance of these processes.

Evapoconcentration is a process in which a fraction of a water volume is evapotranspired and the remaining fraction gets more concentrated, as it holds the total pollutant (normally salt) load of the whole initial water volume. The salinity of irrigation water is the most important variable from the point of view of this process. If salinity is high, the higher consumed fraction in the modernized areas could increase soil solution salinity to levels harmful for crop production. In addition, this increase in soil solution salinity may lead to the precipitation of some low soluble calcium minerals, thereby increasing soil solution sodicity and the corresponding negative effects on the structural stability of soils. This effect is also potentially negative for users who directly use the irrigation return flows.

The salinity of irrigation water is generally low in the ERB (usually < 0.4 dS/m at 25°C) (Isidoro & Aragüés, 2007). Hence, the impact of the evapoconcentration process following irrigation modernization will also be low. This fact will allow optimum on-farm irrigation efficiencies because large leaching fractions will not be required for a proper control of the potential negative on-site effects (i.e., soil salinization and sodification).

Weathering is a process in which a mass of a solid is added to a given water volume. Switching from surface to pressurized irrigation reduces the weathering process because lower percolation volumes decrease mineral dissolution and, depending on the leaching fraction, may promote mineral precipitation. Moreover, fertigation applied with pressurized systems might improve fertilization scheduling and reduce fertilizer doses in relation to old surface irrigation systems, therefore decreasing the input effect.

Table 1 shows the salt and nitrate concentrations and loads measured in the irrigation return flows (IRF) of six study areas located in the central ERB. The pressurized irrigated areas present lower return water volumes and lower salt and nitrate loads than the surface irrigated areas. Salt loads mainly depend on soil salinity and the salinity of underlying geological strata (the two sprinkler areas were high in geologic salts), whereas nitrate loads are much lower in high-efficiency sprinkler irrigated areas than in low-efficiency surface irrigated areas.

The intensification of crop patterns may increase the total amount of agrochemicals and, in particular, fertilizers applied in an irrigation area, therefore increasing the input effect. Hence, if fertilization management is inadequate, the mass of fertilizers exported in drainage water could increase. Recent experiences in sprinkler irrigated areas of the ERB show that, in general, this has not been the case (Cavero *et al.*, 2003; Tedeschi *et al.*, 2001). Similar consequences may occur in the case of other agrochemicals, although the most dangerous, slow-degrading herbicides and pesticides are being progressively banned by European Union regulations.

In the ERB, as in most arid and semiarid inland irrigated areas, the weathering process is more important than the evapoconcentration process because of the vast presence of minerals in the soil or subsoil (Herrero & Snyder, 1997). Since the volume of return flows will decrease, the salt load will be reduced by the modernization process, and the quality (i.e., dissolved solids concentration) of the receiving water bodies will be improved.

These changes in volume and pollutant loads of IRF will also lead to increases in the concentration of these pollutants in the return flows (Table 1). Therefore, after modernization the direct users of these return flows will have reduced water availability, both in terms of quantity and quality. If these return flows are the only water available for these users, other water sources should be allocated to them to compensate for decreased water availability.

Policy Implications

The findings obtained in this work indicate that the irrigation modernization policy set up by the government of Spain will effectively enhance the economic, social and environmental conditions of the agricultural sector. Competitiveness will improve due to an enhancement of irrigation infrastructures. The drudgery of farmers will be greatly ameliorated and these jobs will be more attractive to young rural people. Socio-economic development will be strengthened in rural regions, contributing to maintaining a population balance among regions. Reductions in the export of pollutant loads from irrigated areas will result in significant improvements in water quality, thus helping to achieve the environmental goals set by the European Water Framework Directive.

Irrigation modernization will not, however, result in water savings at the scale of the basin. In cases such as the ERB, where most of the return flows can be re-used, the increase in crop yields will lead to an increase in evapotranspiration and therefore in water depletion. This fact will increase competition for water resources where water is scarce. As a consequence, irrigation modernization will not prevent or solve water conflicts.

Irrigation modernization represents a step forward in the process of conversion from protected to competitive agriculture. In the context of growing demand for food for an

Table 1. Volume, total dissolved solids (TDS), nitrate concentration (NO_3^-), salt and nitrate loads and nitrate losses (as % of N fertilization) in the irrigation return flows of several irrigated areas in the central Ebro river basin

Zone	Area ha	Soil salinity EC_e dS m^{-1}	Annual irrigation return flows					NO_3^- losses %
			Volume mm	TDS mg l^{-1}	NO_3^- mg l^{-1}	Salt load t ha^{-1}	Nitrate load kg NO_3^- ha^{-1}	
SURFACE IRRIGATION								
Bardenas A	95	<2.0	755	541	58	4	98	44
Bardenas B	216	<2.0	1,113	423	77	5	195	56
Bardenas C	409	2.0–6.0	635	2,170	–	14	–	–
Violada	3,866	2.2 (gypsum)	989	1,751	28	20	83	26
Average			958	1,702	31	19	89	28
SPRINKLER IRRIGATION								
Monegros II A	494	2.0–6.0	48	–	125	–	14	10
Monegros II B	470	2.0–6.0	194	6,983	112	14	49	22
Average			119	6,983	119	14	31	16

Note: The electrical conductivity of the irrigation water is low (<0.4 dS m^{-1} at 25°C in all areas).
Source: Basso, 1994; Causapé *et al.*, 2004a; Causapé *et al.*, 2004b; Cavero *et al.*, 2003; Isidoro *et al.*, 2006a; Isidoro *et al.*, 2006b; Tedeschi *et al.*, 2001.

increasing world population, open markets involve an intensification of resource use. If resources such as water and land are limited, their productivity must be increased. Achieving high productivity requires technology, but also knowledge and skills. Therefore, modernization of water management is required in addition to the modernization of irrigation structures.

Modernizing water management requires addressing the three levels of responsibility: farmers, irrigation districts, and river basin authorities. Following modernization, farms will be equipped with new irrigation systems and electronic controllers capable of applying an advanced irrigation scheduling. Farmers must learn how to manage these systems to obtain the maximum benefit from water, fertilizers and other inputs (Tedeschi *et al.*, 2001; Cavero *et al.*, 2003; Dechmi *et al.*, 2003). In the context of the highly variable prices of crops, increased energy prices, and water scarcity, this task will become increasingly complex.

This complexity will lead farmers to request more services from irrigation districts. The districts will have to optimize the ability of their new pressurized water conveyance networks to provide value-added services supporting farmers to achieve their goals. Collective remote control and monitoring systems are being installed in these networks as part of the modernization process. Simulation models and benchmarking techniques must be applied by specialized district personnel to obtain the best performance from these new and complex structures (Zapata *et al.*, 2009).

River basin authorities should provide adequate conditions for optimizing socio-economic returns from water use, achieving the good environmental status of all water bodies, and preventing social conflicts. Development of water accounting concepts will be required, particularly in basins or sub-basins affected by water scarcity. Increased efforts in water-use data collection (identifying the destinations of the water used) will be required, so that accurate water balances can be developed on a routine basis.

Water management modernization will probably require a modification of the relevant legal framework. The core of this legislation was developed many years ago in different socio-economic circumstances when scarcity was due mainly to the lack of technology and investment in developing water resources. Under current water laws, rights are based on water withdrawals. There are no clear distinctions between water use and water depletion, and the efficiency concept is applied to estimate irrigation water demand. Where water is scarce, computation of water withdrawal instead of water depletion could lead to misunderstandings, undermining the decision-making process and the required transparency—key factors in an integrated water management system.

Moreover, from a water quality perspective, policy and legislation will have to be modified or expanded to take into account water pollution problems originating from irrigated agriculture. Currently, the European Water Framework sets limits on pollutant concentrations in order to achieve a 'good environmental status' of water bodies. However, total maximum daily loads that can be exported to each water body are not determined. This approach would promote improvements in water, fertilizer and agrochemical management.

Water management modernization will be much cheaper than infrastructure modernization. However, increased efforts in training and research will be required at all levels: farms, irrigation districts and river basin authorities. The government should promote this process because it will be essential to achieving the maximum economic, social and environmental returns from ongoing public and private investments in irrigation modernization.

Conclusions

The irrigation modernization process linked to an increase in land productivity implies an increase in beneficial evapotranspiration. If the location of the irrigated areas and the quality of the irrigation return flows allow their re-use, as in the Ebro river basin, this increase in beneficial evapotranspiration results in a net increase in water depletion. In these cases, non-beneficial evapotranspiration and non-reusable runoff/percolation are low. Only sprinkler-irrigated areas present sizeable non-beneficial evapotranspiration due to WDEL, contributing to increased water depletion when modernization prescribes switching from surface to sprinkler irrigation. Therefore, water availability will be reduced, owing to irrigation modernization, and despite the increase in on-farm irrigation efficiency.

The magnitude of the increase in the beneficial evapotranspiration will depend on the previous conditions of the traditional surface irrigation systems. Additionally, crop and energy prices, water availability, and the competitiveness of farms will influence future trends in water depletion.

Irrigation modernization will improve the quality of water bodies in semi-arid irrigated areas. Increasing the consumed fraction of water use will reduce the volume of return flows and pollutant loads from irrigated areas because of the reductions in water percolation. However, following modernization, direct users of return flows will have access to less water, but with higher concentrations of pollutants.

The importance of these effects on water quality heavily relies on the adequate management of water and fertilizers, as well as on the salinity of soils, geological strata, and irrigation water. New high-tech irrigation structures will facilitate this management.

Water accounting should be applied instead of the irrigation efficiency approach for a clear understanding of the impact of irrigation in the hydrology of a basin. Complete development of this methodology requires increased efforts in water use and water and soil quality monitoring. A water management modernization plan and changes in some aspects of the current water legislation will be required to achieve this goal. The conceptual application of water accounting in the Ebro river basin has shown that if water savings are intended, efforts should focus on evapotranspiration management, rather than only on farm irrigation efficiency.

Acknowledgements

Thanks are due to Silvia Lozano and Idoia Ederra (Riegos de Navarra S.A.), Segundo Herráez and Eva Díaz de Arcaya (Diputación Foral de Álava), Antonio Enjuanes and Joan Anglés (Generalitat de Cataluña) and Miguel Ángel Monsalve (Ager ingenieros irc S.L.) for supplying and processing part of the data used in this study. Thanks are also due to the rest of the researchers of the authors' departments for the review of a preliminary version of this study.

References

Albiac, J., Playan, E. & Martínez, Y. (2007) Instruments for water quantity and quality management in the agriculture of Aragon, *International Journal of Water Resources Development*, 23(1), pp. 147–164.

Allen, R. G., Clemmens, A. J., Burt, C. M., Solomon, K. & O'Halloran, T. (2005) Prediction accuracy for projectwide evapotranspiration using crop coefficients and reference evapotranspiration, *Journal of Irrigation and Drainage Engineering*, 131(1), pp. 24–36.

Allen, R. G., Pereira, L. S., Raes, D. & Smith, M. (1998) *Crop Evapotranspiration: Guidelines for Computing Crop Water Requirements* FAO Irrigation and Drainage Paper No. 56, Rome: FAO.

Aragüés, R. & Tanji, K. K. (2003) Water quality of irrigation return flows, in: A. Stewart & T. A. Howell (Eds) *Encyclopedia of Water Science*, pp. 502–506 (New York, NY: Marcel Dekker).

Basso, L. (1994) *Los retornos salinos del polígono de riego Bardenas I y su contribución a la salinización de los ríos Arba y Riguel [The saline return flows of the Bardenas I irrigation project and their contribution to the salinization of the River Arba and the River Riguel]*, Doctoral dissertation, Universidad de Zaragoza, 1994.

Beguería, S., López-Moreno, J. I., Lorente, A., Seeger, M. & García-Ruiz, J. M. (2003) Assessing the effect of climate oscillations and land-use changes on streamflow in the Central Spanish Pyrenees, *Ambio*, 32(4), pp. 283–286.

Burt, C. M., Clemmens, A. J., Strelkoff, T. S., Solomon, K. H., Bliesner, R. D., Hardy, L. A., Howell, T. A. & Eisenhauer, D. E. (1997) Irrigation performance measures: efficiency and uniformity, *Journal of Irrigation and Drainage Engineering*, 123(6), pp. 423–442.

Castañeda, C. & García-Vera, M. A. (2008) Water balance in the playa-lakes of an arid environment, Monegros, NE Spain, *Hydrogeology Journal*, 16(1), pp. 87–102.

Causapé, J., Quílez, D. & Aragüés, R. (2004a) Assessment of irrigation and environmental quality at the hydrological basin level: I. irrigation quality, *Agricultural Water Management*, 70(3), pp. 195–209.

Causapé, J., Quílez, D. & Aragüés, R. (2004b) Assessment of irrigation and environmental quality at the hydrological basin level: II. salt and nitrate loads in irrigation return flows, *Agricultural Water Management*, 70(3), pp. 211–228.

Causapé, J., Quílez, D. & Aragüés, R. (2006) Irrigation efficiency and quality of irrigation return flows in the Ebro river basin: an overview, *Environmental Monitoring and Assessment*, 117(1–3), pp. 451–461.

Causapé, J. (2009) Agro-environmental evaluation of irrigation land: I. water use in Bardenas irrigation district (Spain), *Agricultural Water Management*, 96(2), pp. 179–187.

Cavero, J., Beltrán, A. & Aragües, R. (2003) Nitrate exported in the drainage water of two sprinkler irrigated watersheds, *Journal of Environmental Quality*, 32, pp. 916–926.

CHE (Confederación Hidrográfica del Ebro) (2005) *Caracterización de la demarcación y registro de zonas protegidas, implantación de la Directiva Marco del Agua, demarcación hidrográfica del Ebro [Characterization of the demarcation and register of the protected areas, implementation of the Water Framework Directive, Ebro river demarcation]* (Zaragoza: Confederación Hidrográfica del Ebro).

Clemmens, A. J., Allen, R. G. & Burt, C. M. (2008) Technical concepts related to conservation of irrigation and rainwater in agricultural systems, *Water Resources Research*, 44, pp. 1–16.

Clemmens, A. J. & Dedrick, A. R. (1994) Irrigation techniques and evaluations, in: K. K. Tanji & B. Yaron (Eds) *Advanced Series in Agricultural Sciences* (Berlin: Springer-Verlag).

Dechmi, F., Playán, E., Faci, J. M., Tejero, M. & Bercero, A. (2003) Analysis of an irrigation district in northeastern Spain: II. irrigation evaluation, simulation and scheduling, *Agricultural Water Management*, 61, pp. 93–109.

European Union (2000) Directive 2000/60/EC of the European Parliament and of the Council establishing a framework for the Community action in the field of water policy, *Official Journal*, L327, 22/12/2000.

Faci, J. M., Bensaci, A., Slatni, A. & Playán, E. (2000) A case study for irrigation modernisation: I. characterisation of the district and analysis of water delivery records, *Agricultural Water Management*, 42, pp. 313–334.

Forteza del Rey, V. (2002) El Plan Nacional de Regadíos, *Agricultura*, 842, pp. 554–556.

Hargreaves, G. H. & Samani, Z. A. (1985) Reference crop evapotranspiration from temperature, *Applied Engineering in Agriculture*, 1(2), pp. 96–99.

Herrero, J. & Snyder, R. L. (1997) Aridity and irrigation in Aragon, Spain, *Journal of Arid Environments*, 35(3), pp. 535–547.

IBRD (International Bank for Reconstruction and Development) (2009) *Global Economic Prospects 2009: Commodities at the Crossroads* (Washington, DC: The World Bank).

INE (Instituto Nacional de Estadística) (1999) *Censo agrario [Census of agriculture]* (Madrid: INE). Available at http://www.ine.es/inebmenu/mnu_agricultura.htm#1 (accessed November 2007).

Isidoro, D. & Aragüés, R. (2007) River water quality and irrigated agriculture in the Ebro basin: an overview, *International Journal of Water Resources Development*, 23(1), pp. 91–106.

Isidoro, D., Quílez, D. & Aragüés, R. (2004) Water balance and irrigation performance analysis: La Violada irrigation district (Spain) as a case study, *Agricultural Water Management*, 64(2), pp. 123–142.

Isidoro, D., Quílez, D. & Aragüés, R. (2006a) Environmental impact of irrigation in La Violada district (Spain): I. salt export patterns, *Journal of Environmental Quality*, 35(3), pp. 766–775.

Isidoro, D., Quílez, D. & Aragüés, R. (2006b) Environmental impact of irrigation in La Violada district (Spain): II. nitrogen fertilization and nitrate export patterns in drainage water, *Journal of Environmental Quality*, 35(3), pp. 776–785.

Jensen, M. E. (2007) Beyond irrigation efficiency, *Irrigation Science*, 25(3), pp. 233–245.

Keller, A., Keller, J. & Seckler, D. (1996) *Integrated Water Resource Systems: Theory and Policy Implications* Research Report 3, Colombo: International Water Management Institute.

Krinner, W., Garcia, A. & Estrada, F. (1994) Method for estimating efficiency in Spanish irrigation systems, *Journal of Irrigation and Drainage Engineering*, 120(5), pp. 979–987.

Lecina, S., Monserrat, J. & Cots, L. (2000) El uso del agua en el regadío de Flumen (II): zona noroccidental de la Acequia de Rufas (Huesca) [The water use in the Flumen irrigated area (II): north-western area of the Rufas' Ditch (Huesca)], *Riegos y Drenajes XXI*, 112, pp. 39–41.

Lecina, S. & Playán, E. (2006a) A model for the simulation of water flows in irrigation districts: II. application, *Journal of Irrigation and Drainage Engineering*, 132(4), pp. 322–331.

Lecina, S. & Playán, E. (2006b) A model for the simulation of water flows in irrigation districts: I. description, *Journal of Irrigation and Drainage Engineering*, 132(4), pp. 310–321.

Lecina, S., Playán, E., Isidoro, D., Dechmi, F., Causapé, J. & Faci, J. M. (2005) Irrigation evaluation and simulation at the irrigation district V of Bardenas (Spain), *Agricultural Water Management*, 73, pp. 223–245.

López-Moreno, J. I., Beniston, M. & García-Ruiz, J. M. (2008) Environmental change and water management in the Pyrenees: facts and future perspectives for Mediterranean mountains, *Global and Planetary Change*, 61(3–4), pp. 300–312.

MARM (Ministerio de Medio Ambiente y Medio Rural y Marino) (2000) *Libro blanco del agua en España [White book of the water in Spain]* (Madrid: MARM).

MARM (2002) *Plan Nacional de Regadíos [The national irrigation plan]* (Madrid: MARM). Available at http://www.mapa.es/es/desarrollo/pags/pnr/principal.htm (accessed: November 2007).

MARM (2004) *Análisis de la economía de los sistemas de producción: resultados técnico-económicos de explotaciones agrícolas de Aragón en 2003 [Economic analysis of production systems: economic-technical results of farms in Aragon in 2003]* (Madrid: MARM).

MARM (2005) *Análisis de la economía de los sistemas de producción: resultados técnico-económicos de explotaciones agrícolas de Aragón en 2004 [Economic analysis of production systems: economic-technical results of farms in Aragon in 2004]* (Madrid: MARM).

MARM (2006) *Plan de choque de modernización de regadíos [Irrigation modernization shock plan]* (Madrid: MARM). Available at http://www.plandechoque-ahorrodeagua.es (accessed November 2007).

MARM (2007) *Encuesta sobre superficies y rendimientos de cultivos:informe sobre regadíos en España [Crop area and yield survey: report on irrigation in Spain]* (Madrid: MARM).

Martínez-Cob, A. (2004) *Revisión de las necesidades hídricas netas de los cultivos de la cuenca del Ebro [Crop water requirements review in the Ebro river basin]* (Zaragoza: Confederación Hidrográfica del Ebro).

Martínez-Cob, A., Playán, E., Zapata, N., Cavero, J., Medina, E. T. & Puig, M. (2008) Contribution of evapotranspiration reduction during sprinkler irrigation to application efficiency, *Journal of Irrigation and Drainage Engineering*, 134(6), pp. 745–756.

Mateos, L. (2008) Identifying a new paradigm for assessing irrigation system performance, *Irrigation Science*, 27(1), pp. 25–34.

Ministerio de Fomento (1879) Ley de Aguas de 13 de junio de 1879, *Gaceta de Madrid*, 170, pp. 799–805.

Molden, D., Oweis, T., Steduto, P., Bindraban, P., Hanjra, M. A. & Kijne, J. (2010) Improving agricultural water productivity: between optimism and caution, *Agricultural Water Management*, 97(4), pp. 528–535.

Molden, D. & Sakthivadivel, R. (1999) Water accounting to assess use and productivity of water, *International Journal of Water Resources Development*, 15(1–2), pp. 55–71.

Molden, D. J., Sakthivadivel, R., Perry, J., Fraiture, C. & Kloezen, W. H. (1998) *Indicators for comparing performance of irrigated agricultural systems* Research Report 20, Colombo: International Water Management Institute.

Nogués, J., Herrero, J., Rodríguez-Ochoa, R. & Boixadera, J. (2000) Land evaluation in a salt-affected irrigated district using an index of productive potential, *Environmental Management*, 25(2), pp. 143–152.

OECD (Organisation for Economic Co-operation and Development) & FAO (Food and Agricultural Organization of the United Nations) (2009) *Agricultural Outlook 2009–2018: Highlights* (Paris: OECD).

Ongley, E. D. (1996) *Control of Water Pollution from Agriculture* FAO Irrigation and Drainage Paper, 5, Rome: FAO.

Perry, C. J. (1999) The IWMI water resources paradigm: definitions and implications, *Agricultural Water Management*, 40(1), pp. 45–50.

Perry, C. (2007) Efficient irrigation; inefficient communication; flawed recommendations, *Irrigation and Drainage*, 56(4), pp. 367–378.

Perry, C., Steduto, P., Allen, R. G. & Burt, C. M. (2009) Increasing productivity in irrigated agriculture: agronomic constraints and hydrological realities, *Agricultural Water Management*, 96(11), pp. 1517–1524.

Playán, E., Cavero, J., Mantero, I., Salvador, R., Lecina, S., Faci, J. M., Andrés, J., Salvador, V., Cardeña, G., Ramón, S., Lacueva, J. L., Tejero, M., Ferri, J. & Martínez-Cob, A. (2007) A database program for enhancing irrigation district management in the Ebro Valley (Spain), *Agricultural Water Management*, 87(2), pp. 209–216.

Playán, E. & Mateos, L. (2006) Modernization and optimization of irrigation systems to increase water productivity, *Agricultural Water Management*, 80(1–3), pp. 100–116.

Playán, E., Salvador, R., Faci, J. M., Zapata, N., Martínez-Cob, A. & Sánchez, I. (2005) Day and night wind drift and evaporation losses in sprinkler solid-sets and moving laterals, *Agricultural Water Management*, 76(3), pp. 139–159.

Playán, E., Slatni, A., Castillo, R. & Faci, J. M. (2000) A case study for irrigation modernisation: II. scenario analysis, *Agricultural Water Management*, 42, pp. 335–354.

Seckler, D. (1996) *The New Era of Water Resources Management: From "Dry" to "Wet" Water Savings* Research Report 1, Colombo: International Water Management Institute.

Seckler, D., Molden, D. & Sakthivadivel, R. (2003) The concept of efficiency in water-resources management and policy, in: W. Kijne, R. Barkers & D. Molden (Eds) *Water Productivity in Agriculture: Limits and Opportunities for Improvement*, pp. 37–51 (Wallingford: CAB International).

Steduto, P., Hsiao, T. C. & Fereres, E. (2007) On the conservative behaviour of biomass water productivity, *Irrigation Science*, 25(3), pp. 189–207.

Tedeschi, A., Beltran, A. & Aragüés, R. (2001) Irrigation management and hydrosalinity balance in a semi-arid area of the middle Ebro river basin (Spain), *Agricultural Water Management*, 49(1), pp. 31–50.

Turral, H., Svendsen, M. & Faures, J. M. (2010) Investing in irrigation: reviewing the past and looking to the future, *Agricultural Water Management*, 97(4), pp. 551–560.

Willardson, L. S., Allen, R. G. & Frederiksen, H. D. (1994) Elimination of irrigation efficiencies. Paper presented at the 13th Technical Conference, USCID (United States Committee on Irrigation and Drainage), Denver, CO, 19–22 October 1994.

Zapata, N., Playán, E., Martínez-Cob, A., Sánchez, I., Faci, J. M. & Lecina, S. (2007) From on-farm solid-set sprinkler irrigation design to collective irrigation network design in windy areas, *Agricultural Water Management*, 87(2), pp. 187–199.

Zapata, N., Playán, E., Skhiri, A. & Burguete, J. (2009) Simulation of a collective solid-set sprinkler irrigation controller for optimum water productivity, *Journal of Irrigation Drainage and Engineering*, 135(1), pp. 13–24.

A Spatial Planning Perspective for Measures Concerning Flood Risk Management

J. M. M. NEUVEL & W. VAN DER KNAAP
Spatial Planning Group, Wageningen University, Wageningen, The Netherlands

ABSTRACT *As a reaction to flooding events, various governments in Europe addressed the need to create more physical space for water. Experiences in the Netherlands have shown, however, that the development and implementation of these measures can result in local opposition. Based on an examination of such conflicts, it is argued that spatial planning should not only be regarded as an instrument for regulating the land required for flood reduction, but also as an important substantive perspective through which participation can be facilitated and through which water management objectives can be balanced with other spatial claims on the landscape.*

Introduction

In the 1990s, some areas in the Netherlands experienced local flooding or were threatened with flooding near rivers. These events triggered a new national water management strategy, which was consistent with other European countries (Bohm *et al.*, 2004; Howe & White, 2004; Johnson & Priest, 2008), and which addressed the need for more physical space for water. Physical measures aimed at providing more space for water to reduce flood risks, are referred to as spatial measures for flood risk management. Examples of such measures to reduce peak water levels in the rivers are the removal of obstacles from the floodplains or the construction of bypasses and secondary channels to circumvent urban bottlenecks. Examples of spatial measures to prevent local flooding are the development of local and regional water retention areas or canal broadening.

The organization of Dutch water management practices is extensively described in the literature (Wiering & Arts, 2006; Wolsink, 2006). Relevant aspects of the current system for the implementation of spatial measures for flood risk management are the strictly functional responsibilities of water authorities for either flood defence, water quality or water quantity management, and the hierarchical organizational culture within these authorities. Water managers are accustomed to operate in a rather autonomous, isolated and technocratic way. Spatial measures for flood risk management, for example, are often developed in the domain of water management and from a water-management perspective. Consequently, other regional and local ideas on both plan development procedures and substantive aspects of the plan are neglected. As a result, policy makers encounter strong local opposition, which sometimes results in major project delays, inadequate solutions or even in the cancellation of projects aimed at providing more space for water (Davidse, 2008; Grijzen, 2008; Neuvel, 2004; Roth *et al.*, 2006; Schuwer, 2008).

In this light, it is argued that a spatial planning approach towards the development and implementation of spatial measures can have an added value in dealing with current and potential conflicts. Therefore, the central questions addressed in this paper are: what types of conflicts appear during the development and implementation of spatial measures for flood risk reduction; and what can be the added value of a spatial planning perspective in dealing with these conflicts?

These research questions are answered through studying four Dutch projects in which spatial measures for the reduction of flood risks were developed. These projects are discussed from a spatial planning perspective. First we define what a spatial planning perspective is, and then we translate this perspective into an operational analytical framework. This framework is used to analyse the case study projects and to increase understanding of practices and conflicts related to the development and implementation of spatial measures. In addition, the spatial planning perspective is used to provide suggestions on how things might be done differently.

Spatial Planning Perspective

Spatial planning focuses on the physical landscape and the activities that take place in the landscape. Based on the demands and requirements of society, spatial planning tries to position activities such as agriculture, recreation or industry, as well as physical structures such as roads or houses, in the landscape (Tewdwr-Jones, 2001). Spatial planning can be perceived as a policy-making process through which the actors involved try to define and create desired spatial situations while defining and preventing undesired ones (Van Leeuwen *et al.*, 2007). Consequently, spatial planning can be characterized as a future-oriented activity.

Similar objectives can also be encountered in transport planning, flood risk management or urban planning. Spatial planning, however, distinguishes itself from these fields through its holistic, integrative approach; it tries to coordinate different spatial sector claims. Spatial planning, for example, tries to balance and, if possible, integrate spatial claims for water retention with claims for residential development. With an interdisciplinary and comprehensive approach, spatial planning strives for a physical organization of space, which is coherent and desirable from both sector and multi-sector points of view.

In addition to these substantive questions, the term ‘spatial’ also links to procedural aspects, for example, about how spatial planning processes should be designed, how spatial planning decisions should come about or how conflicts about both substantive and procedural aspects should be managed. This includes questions about the involvement of local actors and their role in the policy-making process. Of course, substantive and procedural aspects are interrelated. The actors involved, for example, greatly influence the way a problem is defined. Within flood risk management, however, spatial planning is often narrowed down to a regulatory instrument through which land use change in flood-prone areas can be regulated (Hutter, 2007).

Research Approach

Based on ideas of how people perceive reality (Kolkman, 2005; Schön & Rein, 1994; Te Velde *et al.*, 2002), it is assumed that spatial claims and ideas about desired current and future landscapes are influenced by specific convictions, values, norms, interests and knowledge (Table 1). In this respect, our analysis focuses not only on conflicting spatial claims or conflicting ideas regarding current and desired landscapes but also on diverse and often conflicting interests, norms, values, convictions and knowledge.

To characterize and analyse the procedural aspect of policy-making processes and to characterize how water management authorities dealt with local actors and local opposition, theories on interactive policy making were used. Interactive policy making is a popular approach to solve conflicts in policy-making processes and to increase the support of local stakeholders, among others (Edelenbos & Klijn, 2005). It can be regarded as a more collaborative policy-making approach. Interactive policy making approaches are becoming more prominent in planning practices (Albrechts *et al.*, 2001; Edelenbos & Klijn, 2005; Hajer & Zonneveld, 2000). This is why this perspective was chosen to analyse the procedural aspects of the policy-making process.

Different types of interactive policy-making can be characterized by the degree to which people are offered the opportunity of participating in and determining the final outcome of the policy-making process. In addition, different styles of governance can be distinguished in the way the governmental body deals with the target groups of the proposed policy, the stakeholders and ultimately other initiators of spatial measures (Pröpper & Steenbeek, 1999). Based on the level of participation and the governance style, interactive policy-making processes can be characterized by a participation ladder (Table 2).

Governance styles and levels of participation, however, should not be regarded as static, but rather dynamic. They can change over time and coexist alongside one another. Participation can be defined as consulting during the elaboration of a solution or consulting on specific issues, but can be very limited during the process of agenda setting.

Table 1. Factors that influence desired landscapes

Factor	Definition	Examples
Convictions	Convictions are opinions about the way things are. They are strong beliefs that are not easily called into doubt.	The conviction that one should not kill, or that planet earth is a globe.
Values	Values can be described as worldviews about the way things should be. Values are ethically loaded.	Democracy, social justice, solidarity and equity.
Norms	Norms are the translation of values into tangible rules for behaviour.	Equal risk acceptability criteria, traffic rules.
Interests	Interests can be described as ideas about what is perceived as advantageous for individuals or groups, and is therefore something to be striven for. Interests are induced by convictions and values.	Financial interests: can my company still make a profit? Social interests: can my children grow up in a safe environment?
Knowledge	Knowledge is constructed from experiences, facts, stories and impressions. Knowledge is not only the result of observation, but it also steers observation.	Knowledge about risk, knowledge about the viewpoint and behaviour of other actors.

Table 2. Characterization of the participation of local stakeholders

Level of participation	Definition	Governance style
Co-deciding	The development and policy-making process is left to the initiators and stakeholders. This process is facilitated by authorities.	Facilitating
Co-producing	The government authorities and stakeholders are partners. Agenda setting and problem-solving are joint activities. The government commits itself to the results of the process.	Cooperating
Delegating	The authorities give participants the power to make decisions within the decision-making context provided by these authorities.	Delegating
Advising	Citizens have the opportunity to bring up problems and formulate solutions, but the government defines the agenda. The politicians are committed to the results in principle, but can deviate from them.	Participatory
Consulting	Authorities define the problems and possible solutions. Citizens are consulted about the problems and solutions and can give their points of view. The authorities are free to commit to these points of view.	Consultative
Informing	The citizens are informed about the policy process, but are not allowed to have input.	Open-authoritarian
No-participation	The public is not informed about the policy-making process and is not allowed to influence it.	Closed-authoritarian

Therefore, the level of participation and the governance style should be studied over a longer period of time.

The analysis of conflicts, degree of participation and governance styles was done by examining four 'making space for water' (MSW) projects (Table 3, Figure. 1). Within two of the selected cases, there was strong local opposition, ultimately resulting in the cancellation of the projects. The other two projects are still in progress and have some degree of local support. These projects (with their contrasting processes) were selected to compare and understand conflicts that appear in MSW projects.

Table 3. Overview of the projects

Project	Objective	Status	Sources
Schardammerkoog	Water retention area	Cancelled	Neuvel, 2004
Nieuwe Driemanspolder	Water retention area	In progress	Neuvel, 2004
Ooijpolder	Calamity polder	Cancelled	Davidse, 2008; Roth & Warner, 2007; Roth *et al.*, 2006; Warner, 2008
Kampen Bypass	River bypass	In progress	Grijzen, 2008, 2010; Schuwer, 2008

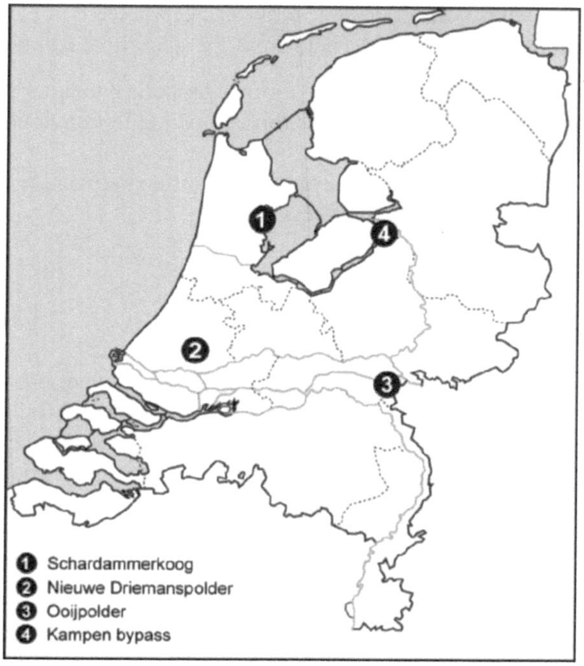

Figure 1. Location map of the projects studied

The analytical framework described above was first used for the examination of conflicts that appeared during two projects for the development and implementation of water retention areas. The research began with a document analysis of scientific literature, policy documents, reports, minutes, newspaper articles and other relevant documents. In addition, key actors from both governmental and non-governmental organizations such as water boards, municipalities, farmers' groups, and inhabitant and nature conservationist groups were selected and interviewed. The interviews were semi-structured with open-ended questions.

In this paper, we compared the results of this research with experiences from two other studies on MSW projects in the Netherlands. These studies focused on the development of a 'calamity polder' (Davidse, 2008) and a river bypass (Schuwer, 2008). The studies were carried out as masters thesis projects, supervised by the present authors. Each piece of research focused on conflicting spatial claims and the arguments on which the actors involved based their claims. In addition, the research focused on the level of participation and governance style, and consequently on the way the policy-making process in these projects came about. This was done using an interactive policy-making framework and methodologies similar to those applied during the analysis of the water retention projects.

The results from these masters theses were augmented with experiences from external studies on the same calamity polder project and the same river bypass project. These studies provided additional information on the projects and the conflicts that appeared during the projects. By bringing these case studies together, the empirical findings provided a more robust starting point for reflecting on conflicts that appeared during these

MSW projects. Subsequently, the four case study projects were compared by asking the following sub-questions (which were derived from the analytical framework):

- Which conflicting spatial claims appeared during the policy-making process and how can these conflicting claims be understood considering convictions, values, norms, interests and knowledge?
- How did the authorities deal with these conflicts in terms of governance style and the level of participation during the process?

Case Studies

Case 1: Water Retention Area—Schardammerkoog

In 1994 and 1998, the north-western part of the Netherlands experienced local flooding. Since studies showed that in the near future, the water system could fail more often due to urban development and climate change, the water board began to search for suitable measures to maintain the current level of protection against local flooding. Next to the water board, the province, a regional agricultural group and a nature conservation organization were involved. This resulted in one preferred solution—a water retention area, to be located in the Schardammerkoog polder, near Lake IJssel. Subsequently, the preferred solution was presented to the local stakeholders in an open authoritarian style. They received a letter in which the proposals for the water retention area were presented.

Both the regional and local actors agreed upon the necessity of additional measures to reduce the probability of local flooding, but they disagreed on the proposed solution. Many local stakeholders still regarded pumping as the best practice to prevent local flooding and questioned the effectiveness of a water retention area. Moreover, a water retention area required 75 hectares of agricultural land and therefore conflicted with agricultural interests. In addition, it conflicted with the desire of the inhabitants to preserve the open landscape. As a result a local pressure group was established representing local farmers and inhabitants during official hearings and meetings with the water board, municipality and province.

This pressure group argued that if water retention areas were needed, several smaller areas in different polders were desirable, rather than one large retention area in their polder. Smaller water retention areas would result in a fairer distribution of the burden. Also the local pressure group heavily criticized the style of governance of the water board. They wanted to be involved at an earlier stage of the policy-making pro-cess.

As a reaction to the local opposition, the water board asked the province to work out the plans in cooperation with local stakeholders, a participatory style. Both local representatives and the province regarded this participation as constructive. Nevertheless, it did not result in the implementation of the water retention area plan. The strong local opposition caused a decrease in the political support for the water retention area, especially amongst the members of the water board council. This council later cancelled the plan for a water retention area.

This project showed how the scope of problems and solutions can become narrowly defined through the involvement of a limited number of actors in the early stage of the policy-making process. Initial proposals for measures were motivated mainly by technical arguments such as flood probabilities, water level reduction and pumping or storage capacity. Through this technical focus and sector-based approach, other values which were based on non-technical or non-scientific arguments or arguments unrelated to water management, such as arguments about distribution of burden, were excluded from the initial debate. This closed style of governance hardly allowed adaptation of the initial plan to local views. Consequently, opportunities to balance the spatial claims of the water authority with the spatial claims of local actors were limited.

The conflicts that appeared in the Schardammerkoog project, however, were not only about the different spatial claims of the actors involved, but also about the values, norms and knowledge underlying the proposal of the water retention area, such as the idea that open landscape should be used for water retention. These values and this knowledge were mainly contested by the actors whose interests conflicted with the proposed spatial measures. Focusing on the values and knowledge underlying the spatial claims instead of the interests involved can be regarded as a way of dealing with the 'dilemma of stake': how to account for interests without being undermined as an interested party (Edwards & Potter, 1992). Affected actors can construct their arguments in such a way that they are not regarded as being simply interest-based. For example, local actors brought the proposed water retention area up for discussion based on the value-rational argument that a large-scale water retention area implies an unfair distribution of burden. Of course, opposing views are not necessarily interest-based. Nevertheless, it can be expected that if proposed spatial measures conflict with interests of local actors, then the underlying values and knowledge on which a measure is based will also be contested.

Case 2: Water Retention Area—Nieuwe Driemanspolder

The south-western part of the Netherlands also experienced local flooding. Consequently, the water board suggested developing a water retention area and increasing the pumping capacity. It was argued that the Nieuwe Driemanspolder would be the most appropriate location for water retention, because this polder was located near the main bottleneck in the water system. In addition, the province already had plans to transform this agricultural area into an area for nature, with agricultural and recreational use, which could be combined with water retention.

Because the province had already started the policy-making process on the coordination of various spatial claims in the Nieuwe Driemanspolder, it became the coordinating authority for the further elaboration of the project. The proposed water retention area lay within the territory of two water boards and three municipalities. For that reason, these authorities formed a steering committee for the project. The objective of this steering committee, however, was not only to develop ideas for the development and implementation of the water retention area, but also to develop a spatial plan for the polder which could balance claims for water retention, open space preservation, agriculture, recreation and nature.

Similar to the Schardammerkoog case, local stakeholders were actively consulted only after the choice for a water retention area in the Nieuwe Driemanspolder was made and elaborated on by the steering committee. The participation of these local stakeholders was formalized through the establishment of a sounding board group in which representatives of inhabitants of the polder, land owners, farmers and nature organizations were involved. The task of this group was to ascertain the desires, ideas and objections of the local stakeholders and to advise the steering committee on the further development and implementation of the water retention area in accordance with a participatory style.

In contrast to the Schardammerkoog case, local pressure groups were not established. Moreover, the proposed water retention area was hardly criticized by local stakeholders. Criticism from local stakeholders focused mainly on the elaboration of the plan rather than on the concept of water retention itself, such as the effect of the actual water quality on the desired natural habitats or compensation for the loss of agricultural land. Inhabitants involved in the sounding board group were concerned about the increase in traffic that was expected from the recreational use of the water retention area. In spite of these concerns, however, the respondents did not reject the water retention plan.

The water retention area seemed to be compatible with the convictions, values, norms, interests and knowledge of local actors representing different sector perspectives such as agriculture, nature conservation or recreation. An important explanation for this 'fit' can be found in the historic development of the policy-making process. In recent decades numerous new functions had been proposed for the polder such as housing development, a rubbish dump, a forest and even a Chinese theme park. The interviewed inhabitants who were members of the sounding board group explained that they were 'not very happy' with the development of a new water retention area in their backyard. The proposed water retention area conflicted with their values, norms and interests because they preferred to keep the area as it was. In spite of this preference, however, they did not take action to prevent the development of a water retention area. They expected that the area would be transformed anyway, and they felt that the development of a water retention area might be 'the best possible development' at the present time and might prevent other, less desirable developments. Farmers in the polder also supported the development of the water retention area. The local farmers wanted to have clarity about their future. The development of a water retention area might be an opportunity for them to sell their land and to develop a new farm in a region with better prospects.

In contrast to the Schardammerkoog project, the Nieuwe Driemanspolder project focused explicitly on balancing different spatial claims. Through this focus, spatial claims for water retention were integrated with claims for agriculture, nature conservation and recreation. Local stakeholders became involved during the elaboration of the plans and supported the proposal for a water retention area. It should be noted, however, that part of the support for the water retention area was actually the response of stakeholders choosing an option they viewed as less threatening than other, less desirable developments, rather than their involvement in the water retention plan itself.

Case 3: Calamity Polder—Ooijpolder

In February 2000, Vice Minister Monique de Vries presented policy proposals for the national policy 'Room for the River', to manage river flooding by using calamity polders (Ministerie van Verkeer en Waterstaat, 2000a). Calamity polders are existing polders that can be deliberately inundated to store water in case of peak water levels. They could also prevent the accidental flooding of other, more densely populated or economically strategic downstream areas. Potential locations for the calamity polders had already been selected in the preparation of the National Spatial Policy Document, using an open authoritarian style. Local and regional authorities and interest groups were surprised by the proposals and heavily criticized the proposed calamity polders. As a reaction to this criticism, a state advisory commission was established by the national government to further investigate the need and conditions for calamity polders.

The Ooijpolder was one of the selected calamity polders. A presentation at the local bank's annual general meeting, in which citizens were informed about the calamity polder plans, caused social turbulence. People were concerned about the consequences of becoming a calamity polder; they were afraid that property values would decline, that desired urban developments would be prohibited and that investments in the area would be reduced (Roth *et al.*, 2006). The presentation triggered local opposition, initially by farmers, but followed by the establishment of a local pressure group. Similar to the Schardammerkoog group, this pressure group included both farmers and a considerable number of inhabitants. The Ooijpolder group, however, lobbied at both the regional and national level, did additional research on the effectiveness of calamity polders, informed the inhabitants of the polder and used the media to increase public awareness regarding their point of view. Similar to the Schardammerkoog group, the arguments used by the local pressure group did not focus explicitly on interests such as property values. The pressure group focused mainly on the knowledge and technical assumptions underlying controlled flooding (Davidse, 2008; Roth & Warner, 2007). They believed, for example, that the estimation used for peak river discharges was too high, undermining the need for calamity polders. This strategy can be regarded as another strategy for dealing with the dilemma of stake—to focus on the assumptions in the plan instead of the interests held by the members of the local pressure group. This prevents the pressure group from acquiring the label of an NIMBY-driven (Not In My Back Yard) pressure group. Consequently, the conflicts were not only about different local and national interests, convictions or values, but also about the technical knowledge on which the proposed calamity polder was based.

In 2003 a consultancy bureau submitted a report to the Ministry of Transport, Public Works and Water Management that criticized the research on calamity polders (Dijkman *et al.*, 2003). The Ministry initially denied the existence of this report towards the local pressure group. The report was only made public after the local pressure group announced it would appeal to the Government Information Act. This and other reports raised growing doubts about the effectiveness of calamity polders and the underlying technical facts (Roth & Warner, 2007).

At this stage, the lobbying of members of Parliament by the local pressure group turned out to be effective as well. One of the contacts of the pressure group helped propose a resolution, which was voted on in Parliament, to transfer the budgeted funds for the calamity polders to the general budget for the Room for the River Policy (Ministerie van Verkeer en Waterstaat, 2000b). The acceptance of this resolution by Parliament forced the national government to postpone the spatial reservation for the calamity polders in the National Spatial Planning Act. This meant that the plans for the calamity polder in the Ooijpolder were removed from the political agenda (Warner, 2008).

As illustrated by the Ooijpolder case study, local actors turned out to be very good at utilizing their networks outside the formal policy-making and information meeting process to influence the outcomes. Experiences in other planning processes have also shown that actors involved in planning continue to communicate via traditional forms and institutions, but that many activities related to spatial planning such as bargaining, decision-making and influencing activities are taking place outside the formal structures and rules of the planning process (Hillier, 2000). Consequently, authorities who are developing spatial water measures that conflict with local ideas on the current and desired future landscape should expect strong local opposition.

Case 4: The Kampen Bypass

In addition to the calamity polders, the national government also considered a plan to develop a river bypass near Kampen, adjacent to the River IJssel to prevent river flooding. The national government put a claim on this area even though the implementation of the bypass was not planned until between 2050 and 2100. This claim conflicted with the housing objectives of the municipality of Kampen, to build 4,000 to 6,000 houses. In addition, the proposed bypass conflicted with other spatial claims such as a new railway and the upgrading of a main road. Therefore, the province brought involved authorities together to develop a spatial master plan for the area and to explore, as had been done in the Nieuwe Driemanspolder project, how the different objectives and claims could be integrated.

The regional master plan was developed in cooperation with local, regional and national authorities. These authorities proposed developing a river bypass in the short term and integrating it with the other spatial claims, such as the development of a railway and the housing plan. Five development scenarios were described in the master plan. They were presented using a consultative governance style to the public by means of three information meetings, open to all interested parties. The participants were allowed to respond to the various scenarios (Grijzen, 2010; Schuwer, 2008).

The scenario for the bypass as a new secondary river, navigable for recreational boating, was most popular with the participants. During the information sessions, however, inhabitants of Kamperveen, a small village near the city of Kampen, were especially opposed to the proposed scenarios because the bypass would physically split their community. In reaction to this resistance, the province offered to support people in developing alternative scenarios with the aid of the experts involved in the development of the master plan. This governance style can be regarded as a more interactive participatory style, since the authorities did not necessarily commit themselves to the result. Through this more participatory style, province representatives successfully transformed local opposition by inhabitants of Kamperveen into constructive participation (Grijzen, 2008). This resulted in a sixth scenario, a variant of the popular secondary river scenario, in which the Kamperveen community was divided as little as possible. Based on the information meetings, the project group opted for the elaboration of a bypass as a secondary river that would be navigable for recreational boating and for a trajectory that did not divide Kamperveen.

Similar to the Nieuwe Driemanspolder project, the elaboration of the master plan was carried out by a project group which consisted of civil servants from the local and regional authorities and a steering group in which the local and regional governments were represented. Representatives of non-governmental stakeholders were not involved in the consultation rounds as they had been during the presentation of the scenarios, but were invited to participate in a sounding board group, without any formal decision-making power.

During the elaboration of the plans, ideas for residential development were worked out further. The proposed residential development in the area, which was needed for financing the desired development, amongst other objectives, was still strongly opposed by some inhabitants. Like the group in the Ooijpolder, the local pressure group from Zwartendijk (another village near Kampen) focused on the knowledge underlying the purported housing need. They argued that the housing need was greatly overestimated and proposed building houses at alternative locations, especially within the city itself (Stichting Werkgroep Zwartendijk, 2007). In addition, some farmers were still dissatisfied with the proposed developments because the plans could result in their relocation. As this article went to press, local farmers and the local pressure group from Zwartendijk are still trying to influence the planning process. Like the Ooijpolder group, they are not only using formal methods of resistance, such as formal appeals, but also more informal ones. Nevertheless, the representatives of local and regional authorities still expect the current plans to be implemented. They regard acquiring sufficient funding for the bypass as one of the main bottlenecks for implementation of the plans (Schuwer, 2008), not the conflicts with local actors.

Similar to the other case study on the development of spatial measures for water management, the proposed bypass near Kampen resulted in conflicts with local actors affected by the bypass. As in the other case study projects, these conflicts were not only about the different interests, but also about the underlying knowledge on which the ideas were based. In the light of the policy-making process, this case study showed how governance styles, and especially adapting governance styles, can contribute to persuading local opposition to engage in more constructive participation (Grijzen, 2008). In addition, the project illustrated that a spatial planning perspective can be used to coordinate various activities that result in spatial claims, such as the development of a bypass, residential areas or new railroad.

Discussion and Conclusions

Spatial measures to reduce flood risks require space, or at least another (combined) use of space. The case studies showed that the spatial claims resulting from the proposed measures conflicted with other spatial claims and underlying values, norms and interests of actors in the area (Table 4). When proposed measures conflict with local views, policy makers can encounter serious opposition organized by the affected local and regional stakeholders. These stakeholders are able to organize strong opposition and make their voices heard. Through this opposition, not only the measures themselves, but also the underlying knowledge and values of the proposed measures are contested. In the end, this can result in the cancellation of projects as illustrated by the Schardammerkoog and Ooijpolder projects.

In this light, we argue that a more interactive approach to policy making could be a way to manage these conflicts to help involved stakeholders recognize their dependency on other actors and learn about the convictions, values, norms, interests and knowledge of other actors involved. Furthermore, more interactive policy making can increase trust amongst involved actors, provide additional, area-specific knowledge and creativity and be helpful in identifying potential conflicting values and interests at an early stage (Innes & Booher, 2003). Nevertheless, the most appropriate level of interaction or governance style cannot be predefined. They are context dependent. Comparable governance styles in the Schardammerkoog and Nieuwe Driemanspolder led to very different outcomes. The Kampen case study, along with previous studies on interactive decision-making processes (Edelenbos & Klijn, 2005), have shown that policy-making processes are promising if levels of participation and styles of governance are adapted to the conflicts at hand, if the interactions are well prepared and if the voiced preferences are taken seriously.

More interactive policy making, however, implies that discussions are broadened to introduce additional local spatial claims, such as agriculture or housing. This study of the MSW projects showed that the proposed measures for flood risk reduction not only

Table 4. Overview of the governance styles and conflicts on plan content

Project	Governance style (from… via… to)	Conflicting points of view
Schardammerkoog	closed authoritarian → open authoritarian → consultative	Knowledge: effectiveness and efficiency of water retention versus pumping Interests: use of land and landscape quality versus water retention Values: distribution of burden Norms: one large or numerous smaller retention areas, involvement of local actors
Nieuwe Driemanspolder	closed authoritarian → open authoritarian → consultative	Knowledge: water quality Interests: continuation of agricultural businesses and livability versus land for water retention and recreation Values: openness versus water safety Norms: open landscape, water retention
Ooijpolder	closed authoritarian → open authoritarian	Knowledge: effectiveness and efficiency of calamity polders with respect to alternative flood mitigation measures Interests: land values, development potential, flood protection Values: protection against flooding, effectiveness, efficiency Norms: level of protection of flood prone areas, effective and efficient measures
Kampen Bypass	open authoritarian → consultative → participatory → consultative	Knowledge: housing needs Interests: river bypass, housing development, infrastructural developments, landscape quality, land ownership, agricultural interests and financial interests Values: openness, peacefulness, social coherence Norms: no bypass through a village, open landscape without houses or extensive recreation

conflicted with local views on water management, but also with local views on the desired current and future land use. For this reason, a spatial planning perspective can be of added value if compared to the sectoral water management perspective. A spatial planning perspective implies a substantive focus on the landscape as a whole and on the coherence of and between different land uses and inhabitants. This perspective emphasizes a more holistic starting point through which spatial measures for water management can be discussed in relation to other spatial claims. Moreover, the proposed measures and underlying arguments can be confronted with other convictions, values, norms and interests with respect to current and future land use. As illustrated by the Nieuwe Driemanspolder and the Kampen Bypass project, for example, a more holistic approach offers the opportunity to balance and integrate spatial claims for water management with other claims such as agriculture, housing or transport.

In the introduction, it was argued that the space for water policies could result in a greater priority for water issues within spatial planning practices (Howe & White, 2004; Wolsink, 2006). These case studies, however, have made it clear that in MSW practices, a greater priority for a spatial planning approach and the consistent consideration of other spatial claims, values and interests—such as landscape preservation or agriculture—are required. Within flood risk management, however, spatial planning is often narrowed down to a regulatory instrument through which, for example, land-use changes in flood-prone areas can be regulated (Hutter, 2007). However, as this paper has illustrated, there is more to spatial planning than that. In addition to providing an interesting research perspective, spatial planning also provides added value as a substantive focus in interactive policy-making processes. This was shown in the development and implementation of spatial measures for water management, since this type of policy making inherently requires balancing different spatial claims with varying interests, values, norms and convictions with respect to the current and future landscape and its inhabitants and users.

Acknowledgements

The authors would like to thank Dries Schuwer and Bart Jan Davidse for their help in data collection and Madelinde Winnubst and Jantine Grijzen for their constructive comments on an earlier version of this paper. This paper was based on research funded by the Dutch innovation programme "Space for Geo-information".

References

Albrechts, L., Alden, J. & Da Rosa Pires, A. (Eds) (2001) *The Changing Institutional Landscape of Planning* (Aldershot: Ashgate).

Bohm, H. R., Haupter, B., Heiland, P. & Dapp, K. (2004) Implementation of flood risk management measures into spatial plans and policies, *River Research and Applications*, 20(3), pp. 255–267.

Davidse, B. J. (2008) Besluitvorming over binnendijkse maatregelen ter bestrijding van overstromingen [Decision making on flood risk management in dike ring areas], Masters of Science thesis, Wageningen University, 2008.

Dijkman, J. P. M., Ogink, H. J. M., Klijn, F. & Van Der Most, H. (2003) *Aanvullend deskundigenoordeel noodoverloopgebieden (Additional Expert Judgement on Calamity Polders)* (Delft: WL Delft Hydraulics).

Edelenbos, J. & Klijn, E. H. (2005) Managing stakeholder involvement in decision making: a comparative analysis of six interactive processes in the Netherlands, *Journal of Public Administration Research and Theory*, 16(3), pp. 417–446.

Edwards, D. & Potter, J. (1992) *Discursive Psychology* (London: Sage Publishers).

Grijzen, J. (2008) A whole village buried alive: the dramaturgy of participation in water projects. Paper presented at the Freude am Fluss Final Conference: Space for the River, Space for People? Nijmegen, October.

Grijzen, J. (2010) *Outsourcing Planning. What Do Consultants Do in a Regional Spatial Planning in the Netherlands* (Amsterdam: Amsterdam University Press).

Hajer, M. & Zonneveld, W. (2000) Spatial planning in the network society: rethinking the principles of planning in the Netherlands, *European Planning Studies*, 8(3), pp. 337–355.

Hillier, J. (2000) Going round the back? Complex networks and informal action in local planning processes, *Environment and Planning A*, 32(1), pp. 33–54.

Howe, J. & White, I. (2004) Like a fish out of water: the relationship between planning and flood risk management in the UK, *Planning, Practice & Research*, 19(4), pp. 415–425.

Hutter, G. (2007) Strategic planning for long-term flood risk management: some suggestions for learning how to make strategy at regional and local level, *International Planning Studies*, 12(3), pp. 273–289.

Innes, J. E. & Booher, D. E. (2003) Collaborative policymaking: governance through dialogue, in: M. A. Hajer & H. Wagenaar (Eds) *Deliberative Policy Analysis: Understanding Governance in the Network Society*, pp. 33–59 (Cambridge: University Press).

Johnson, C. L. & Priest, S. J. (2008) Flood risk management in England: a changing landscape of risk responsibility?, *International Journal of Water Resources Development*, 24(4), pp. 513–525.

Kolkman, M. J. (2005) *Controversies in Water Management: Frames and Mental Models* (Enschede: Febodruk).

Ministerie van Verkeer en Waterstaat (2000a) *Ruimte voor de Rivier* [*Room for the River*] (Den Haag: Ministerie van Verkeer en Waterstaat, Directoraat-Generaal Rijkoswaterstaat).

Ministerie van Verkeer en Waterstaat (2000b) *Planologische Kernbeslissing Ruimte voor de Rivier. deel 4. Vastgesteld besluit en Nota van Toelichting* [*National Spatial Planning Key Decision. Room for the River, part 4. Decision and Commentary*] (Den Haag: Ministerie van Verkeer en Waterstaat).

Neuvel, J. M. M. (2004) *Wateroverlast en watertekort: Percepties op risico's en consequenties voor de ruimtelijke ordening* [*Local flooding and water shortage: risk perceptions and their consequences for spatial planning*], Report for Milieu en Natuurplanbureau, Bilthoven, Netherlands.

Pröpper, I. & Steenbeek, D. (1999) *De aanpak van interactief beleid: Elke situatie is anders* [*The Interactive Policy Approach: Every Situation is Different*] (Bussum: Coutinho).

Roth, D. & Warner, J. (2007) Flood risk, uncertainty and changing river protection policy in the Netherlands: the case of 'calamity polders', *Tijdschrift voor Economische en Sociale Geografie*, 98(4), pp. 519–525.

Roth, D., Warner, J. & Winnubst, M. (2006) *Een noodverband tegen hoog water: Waterkennis, beleid en politiek rondom noodoverloopgebieden* [*A first-aid against high water levels. Water knowledge, policy and politics related to calamity polders*] (Wageningen: Wageningen UR).

Schön, D. & Rein, M. (1994) *Frame Reflection: Towards the Resolution of Intractable Policy Controversies* (New York: Basic Books).

Schuwer, D. (2008) Ruimte voor de IJssel: Een onderzoek naar de nieuwe regionale plannen van Zuthpen en Kampen [Room for the river IJssel: an examination of the proposed regional spatial plans of Zutphen and Kampen], Masters of Science thesis, Wageningen University, 2008.

Stichting Werkgroep Zwartendijk (2007) *Een dijk te ver. Een andere visie op de ontwikkeling van Kampen en IJsseldelta Zuid* [*A dike too far. An alternative perspective on the urban development of Kampen and the IJsseldelta South area*]. Report for Stichting Werkgroep Zwartendijk, Kampen, Netherlands.

Te Velde, H., Aarts, M. N. & Van Woerkum, C. (2002) Dealing with ambivalence: Farmers' and consumers' perceptions of animal welfare in livestock breeding, *Journal of Agricultural & Environmental Ethics*, 15(2), pp. 203–219.

Tewdwr-Jones, M. (2001) Complexity and interdependency in a kaleidoscopic spatial planning landscape for Europe, in: L. Albrechts, J. Alden & A. Da Rosa Pires (Eds) *The Changing Institutional Landscape of Planning*, p. 8–34 (Aldershot: Ashgate).

Van Leeuwen, E. S., Hagens, J. E. & Nijkamp, P. (2007) Multi-agent systems: a tool in spatial planning, *DISP*, 170(3), p. 19.

Warner, J. (2008) Emergency river storage in the Ooijpolder—a bridge too far? Forms of participation in flood preparedness policy, *International Journal of Water Resources Development*, 24(4), pp. 567–582.

Wiering, M. A. & Arts, B. J. M. (2006) Discursive shifts in Dutch river management: 'deep' institutional change or adaptation strategy? *Hydrobiologia*, 565(1), pp. 327–338.

Wolsink, M. (2006) River basin approach and integrated water management: governance pitfalls for the Dutch space-water-adjustment management principle, *Geoforum*, 37(4), pp. 473–487.

Water Governance: Some Critical Issues

CECILIA TORTAJADA
Lee Kuan Yew School of Public Policy, Singapore; and International Centre for Water and Environment (CIAMA), Zaragoza, Spain

ABSTRACT *This paper presents an analysis of the issues discussed at a special international workshop on water governance. While it is generally accepted that good governance for the water sector is essential, it is also clear that its implementation requires qualitative and quantitative factors, which may vary from one country to another. In order to objectively assess the opportunities and constraints of implementing good water governance practices, a group of selected international experts were invited to address this complex issue.*

Background

The International Workshop on Water Governance was organized in June 2009 by the Institute of Water Policy of the Lee Kuan Yew School of Public Policy, National University of Singapore, and the Third World Centre for Water Management in Mexico. This workshop to discuss the critical issues of water governance, and a further event to develop a priority research agenda on the topic of water governance (Tortajada, 2010) were part of the First Global Water Policy Dialogue of the Institute of Water Policy. The annual policy dialogues have the objective to provide a forum for discussions of mainly intersectoral, interdisciplinary future-oriented events on a global basis on new and innovative policy alternatives for efficient water management, including their economic, social, political, institutional, legal and environmental implications.

The workshop included three main presentations to set the scene for discussions on the various issues associated with water governance, and a panel session to identify critical and priority aspects of water governance from different water resource sectors and from diverse disciplines and perspectives. Following the workshop, there was a brainstorming session on water governance. A summary of the discussions of this session is published in this issue.

During the welcoming and opening addresses of the workshop, K. E. Seetharam, Director of the Institute of Water Policy at the Lee Kuan Yew School of Public Policy, and Woochong Um of the Asian Development Bank, stressed the relevance and need for meaningful water policy dialogues in terms of water governance. The main presentations of the workshop included: (1) plans, objectives and outputs of the workshop as well as an overview of the approaches, issues, experiences and shortcomings of water governance by Asit K. Biswas; (2) perspectives on water governance from the agricultural sector by David

Molden; and (3) urban water management by this author. The session was chaired by Alexander Zehnder of the Alberta Water Research Institute in Edmonton, Canada.

The panel session was chaired by Asit K. Biswas, and the panellists included the following eminent personalities: Kishore Mahbubani, Dean, Lee Kuan Yew School of Public Policy, Singapore; Gee Paw Tan, Chairman, Public Utilities Board, Singapore; Alexander Zehnder, President, Alberta Water Research Institute, Canada; Mahmoud Abu-Zeid, President, Arab Water Council, Cairo, Egypt; Dieter Ernst, CEO, Wasser Berlin, Germany; and Gérard Payen, President, Aquafed, France.

The workshop ended with remarks from Asit Biswas on the possible future follow-up activities on water governance based on the discussions and conclusions of this event. Following is the summary of key discussions at the above workshop.

Governance

Governance has been used mostly as an umbrella concept and no agreed definition exists. Governance is not synonymous with government. It is instead a complex process that considers multi-level participation beyond the state, where decision making includes not only public institutions, but also the private sector, civil society and society in general. Good governance frameworks refer to new processes and methods of governing and changed conditions of ordered rule on which the actions and inactions of all parties concerned are transparent and accountable. It embraces the relationships between governments and societies, including laws, regulations, institutions, and formal and informal interactions which affect the ways in which governance systems function, stressing the importance of involving more voices, responsibilities, transparency and accountability of formal and informal organizations associated in any process.

Because of its complexity, good governance clearly does not just appear, but has to be carefully planned. It is the culmination of multifaceted, long-term processes that have to be properly planned and nurtured. For good governance to develop, overall conditions and the general environment must be made favourable: parties concerned must be amenable to collective decision making; effective and functional organizations need to be developed; and policy, legal and political frameworks must be suitable to the goals that are being pursued for the common good (Rhodes, 1996; Kooiman, 2003; Tiihonen, 2004).[1]

Since governance-related issues are not just public or private, but are frequently shared, governance activities at all levels become diffused over various societal actors whose relationships with each other are constantly changing. The challenge for anyone involved in governing and governance is to make public, private and societal actors participate actively and directly at solving problems and creating opportunities under both normative and institutional frameworks that provide the foundations for any activity. For these complex interrelationships to succeed they have to take into consideration that they are interdependent, and that no single actor, public or private, has the knowledge, information or power to solve the changing societal challenges on one's own. This is because no actor has sufficient knowledge or wherewithal to make the necessary instruments effective on his own, or sufficient action potential to unilaterally dominate the decision-making arena (Kooiman, 2003).

Regarding civil society, governance enables it to play a role both as a responsible stakeholder and as an increasingly important force for reforms and development processes. Nevertheless, involving civil society as a stakeholder is a very complex political, philosophical and technical task for institution-building in any country, mainly because of its multidimensional and multi-stakeholder nature. Responsibilities cannot be transferred from the state to society simply because there is no one monolithic group known as 'society.' Society is composed of heterogeneous groups of individuals, citizens, organized associations and unorganized communities with very complex relationships and varied interests. The complexities of these interrelationships between the actors and institutions, and their continuing dynamism and evolution, have meant that governance discourses can mostly be implemented only to a certain extent, and that its application often remains mainly theoretical in nature and concerns mostly the development of strategies and principles, but not so much the difficult execution or implementation phase (Tiihonen, 2004).

The changing dynamics of governance has meant that the identification of the development needs and the implementation of programmes have to take place in the form of partnerships and within the context of the full understanding and appreciation of the intersectoral linkages. Consequently, country development strategies have to search for agreements, commitments and contributions, not only from governments at different levels, but also from domestic and foreign private sectors, development agencies and the stakeholder groups of civil society.

Water Governance: The Concept

The concept of water governance, as is the case for governance in general, is still evolving. At present, there is no universally agreed upon definition for water governance, and its ethical implications and political dimensions are a matter of national and international debate. The result is that different people and institutions are using the concept in different ways, and within varying cultural, economic, social, legal and political contexts.

Water governance can be perceived, in its broadest sense, as comprising all social, political, economic and administrative organizations and institutions, as well as their relationships to water resources development and management. It is concerned with how institutions operate and how regulations affect political actions and societal concerns through formal and informal instruments (UNDESA *et al.*, 2003).

UNDP (2004) considers water governance to include political, economic and social processes and institutions through which governments, private sector and civil society make decisions about how best to use, allocate, develop and manage water resources. It refers to a range of political, social, economic, and administrative systems that are in place to develop and manage water resources and the delivery of water services at different levels of society and for different uses. It comprises mechanisms, processes, and institutions through which all involved stakeholders, including citizens and interest groups, articulate their priorities, exercise their legal rights, meet their obligations and mediate their differences. It emphasises the causality of water-related problems by pointing out not only the natural limitations of the water supply and/or lack of financing and appropriate technologies, but rather from profound failures in water governance, such as the ways in which individuals and societies have assigned value to, have made decisions about, and have managed the water resources available to them.

Water governance and water management are interdependent issues in the sense that effective governance systems are meant to enable practical management tools to be applied properly as situations require. Partnerships between the public and private sectors, participation of stakeholders, and economic or regulatory instruments will not be effective unless there are administrative and management mechanisms in place, as well as commitments and involvements of governments, private sector groups and civil society organizations. Even though reform of water institutions and policies are taking place in many countries, progress in nearly all cases has been somewhat slow and limited, and often unpredictable. In most of the countries of the developing world, water institutions do not function properly, and most of them display fragmented institutional arrangements and overlapping and/or conflicting decision-making structures. Some even consider water governance as the framework within which integrated water resources management can be applied (Rogers & Hall, 2003). While some claim that integrated approaches are of fundamental importance to manage water in more efficient ways, the fact remains that their implementation has remained incomplete and unsatisfactory in all countries, developed or developing, after some 60 years of trying.

It is true that involvement in policy formulation, coordination, communication and information exchange between administrations and stakeholders at various levels would contribute to a better knowledge of each others' policy objectives, working modalities and mechanisms available for implementation. However, decisions should first be coherent with a broad set of principles leading to a more progressive and efficient and equitable management of water resources. For example, it is expected that by directly involving users of water in the governance of the resource, the knowledge on which this participation is based will be more meaningful than otherwise may have been possible since local stakeholders may often be more familiar with the peculiarities of local economic, social, cultural and environmental situations. It might also mean that users involved may accept the regulations as appropriate and consistent with their values and interests and may also be more willing to comply with them (Kooiman, 2003). Nevertheless, in practice rather than in theory, many stakeholders, including water users associations, non-governmental organizations (NGOs) and local communities, may build on local knowledge and networks, but may often lack support, funds, institutional and technical capacities or even membership to significantly contribute to the management of water resources. Therefore, the means for enabling and coordinating the effective roles of different stakeholders in water governance remains an imperative but elusive challenge that has to be solved before the concept of governance can become a reality.

Water management is evidently not an issue that can be approached only from within the water sector, or only from the perspective of one single stakeholder or sector. It is instead a development challenge that requires the cooperation, collaboration and coordination from within and outside the water sector as well as from the multiple interested parties. As important as the meaning of the concept of water governance is, countries and sectors are trying to implement it when they have still not solved some of their long-standing and rampant social and economical problems that have been confronting them for decades. In the water sector, non-functional water institutions, deficient legislations, and overall inefficiency in managing water resources in most developing countries, have been well-known constraints for decades. With the involvement of multiple actors in the arena of water resources planning, development and management, and the introduction of ethical issues such as responsibility, accountability, transparency, equity and fairness, the challenges associated with making good water governance a reality has become even more complex. It now remains to be seen how countries, their governments and their societies can make the best of the opportunity to develop effective partnerships and encourage partner participation, while giving consideration to ethical, social and environmental issues necessary to achieve the much-needed reforms in the water sector.

Seeking improved patterns of governance within an environment of different interests, dissimilar values and norms, when many times there is an absence of consensus even about goals and objectives, represents formidable challenges that require extraordinary measures of coordination, cooperation and compromise not only from the different levels of government, but also from private sectors and social stakeholders in general. Therefore, the way forward for implementation of the concept of good water governance is to realize the importance of planning and implementing frameworks within specific social, economic, environmental and cultural conditions, including processes and mechanisms of interaction between state and non-state actors looking for efficiency and mutual responsibilities.

Water Governance: Key Issues for Implementation

At present, some of the common features required to achieve so-called good governance in the water sector are said to be functional institutions within a multi-sector, multi-level perspective in order to avoid major gaps or overlaps in policy, planning and funding, as well as formulation and implementation of policies, laws and regulations. Nevertheless, management and budgetary practices are still expected to deliver services with tariff structures that balance affordability, equity and sustainability. Proper monitoring, evaluation and reporting at regular intervals are not only yet to be applied to formal institutions, but also to informal institutions, such as non-governmental organizations and citizens groups. Nevertheless, when it comes to ensure that the above tasks are objectively, adequately and equitably performed, the decision of which formal or informal group has the authority and credibility to carry out the corresponding evaluations is still a matter of debate.

In terms of indicators for good governance, these were initially developed with the objective to assess the performance of public and private institutions, primarily from economic and financial viewpoints. However, as important as the concept of good governance is to all development-related activities, the fact that it is both subjective and objective makes exceedingly difficult the development of a meaningful framework which can be implemented. There are too many issues related to governance that remain unresolved and also somewhat vague, and which still require better definition, appreciation and understanding. For example, if participation is important, how do we get informal institutions, including citizens, involved in decision making that renders fruitful results? If the above sectors do get involved, how can we ensure that the outcomes achieved will be useful and better than otherwise may have been the case? Who decides who should be the participants that should be involved? Should they be only from the area where specific water development projects affect their lives, or should they include people from outside the region, or even the country? If people from outside the region, or even the country, are to be considered as stakeholders of a project, what importance should be given to their views compared to the people from the project-affected areas? Unfortunately, debates and discussions on such fundamental and complex issues have not yet been started and are still very far from being answered, the main constraint being that participation is, in most of the cases, considered as an end by itself instead of as a means to an end (Tortajada, 2007).

In addition to participation, accountability and transparency are considered to be other fundamental elements of good governance. However, their complexity is such that fundamental issues, such as how accountability should be approached and to whom the policy makers and/or project authorities should be accountable, are still a matter of debate. In terms of transparency, while there is general consensus that this is an essential component of governance, there is no agreement as to which entity should be most transparent for governance to be credible: government institutions, elected representatives, and/or the general public? One more topic of intense deliberation concerns what process or processes should be instituted to ensure that accountability is genuine and relevant to the appropriate groups, especially as the requirements for transparency are different depending on the groups in question, on their interests and their views.

Agricultural Water Management[2]

In the field of agriculture, a challenge that seriously limits the operationalization of the concept of water governance is the fact that the focus in the water sector is still in the construction of new projects and not yet on the management of the resource. That is, it is still necessary to move from the overall construction-oriented focus to the management perspective. Only afterwards may it be possible to understand the formal and informal nature of the elements of water governance to work towards public policy goals, including its institutions, regulations, partners, and direct and indirect instruments.

Some of the challenges in the agricultural water management sector include:

- the need to work with the informal nature of water economies that already exist in many countries;
- the realization that we have reached the upper limits of water use in many basins of the world and our systems have thus become much less flexible;
- the fact that transboundary water management requires broader views which should include social and environmental concerns; and
- the limitation that results because we continue seeing water in isolation from other policy levers outside the water sector, such as the energy or food sectors, which could provide us with further insights for efficient agricultural water governance.

Offering equal incentives to different people or institutions with different, and even conflicting goals and agendas has repeatedly proved to be a challenge.

Some prescriptions for improved water governance include:

- treating water as an economic good;
- declaring water a state property and creating water rights;
- introducing new implementable water laws and policies;
- establishing effective participatory structures and processes; and
- managing water at the basin level.

Nevertheless, in many countries, especially in Africa, global paradigms may have become a hindrance rather than a solution. Prescriptions for improved governance so far show too little evidence that the accepted paradigms are working over the long-term. Some examples include: the consideration of water as an economic good in Tanzania and Sri Lanka with flat water tax tariffs; water rights in sub-Saharan Africa; overall water management at the basin level; and participatory structures that still have not delivered useful results.

Tried and tested solutions developed for one context may or may not be applicable to other locations simply because technical, social, economic and environmental conditions are very different. They may not even be easily transferable. There are limits to 'best practices' and we may have to start looking for good and implementable solutions such as indirect public policy instruments (e.g., Jyotigram in India), build on success from the informal sector (e.g., Zhang He in China) and promote institutional innovations that rationalize incentives and reduce transaction costs (e.g., Zhang He and Jyotigram). The successful marriage between instruments from the informal and formal sectors are likely to result in the improvement of water infrastructure and services through improved investment patterns and management practices, and also to more formal arrangements when appropriate. Overall, it should be remembered that without good water governance, it will not be possible to achieve water security on a long-term basis in any country.

Urban Water Management

In most cities of the developing world, clean water is both scarce and expensive to produce. Large sectors of the population in densely populated urban areas do not have access to potable water, and even larger sectors of the population do not have access to sanitation. The situation normally includes non-revenue water, often up to 40–60%, infrastructure either scarce or becoming complex and deteriorating, water supplies largely underpriced, and investment needs, possibly reaching billions of dollars. All the above issues are the result of a lack of appropriate water and land resources management, and lack of adequate institutions and policies which address them beyond the short-term. An important constraint is that, in spite of its importance for development, water is not the critical development issue in urban areas. Transportation, energy and housing often take priority for the development of cities.

Urban areas face challenges that require immediate action, and governance has become one of the most pressing ones. In order to be sustainable, urban areas need, in addition to economic growth and implementable social and environmental policies, proper governance systems which include adequate mechanisms for broader and meaningful participation. Since all of these aspects are intertwined with the various cross-cutting roles that water plays in urban development as an important contributor to all the other sectors, there is thus a fundamental reason why the process of water governance has become so complex and critical.

There are proposals for alternatives to improve the governance of urban areas and provide citizens with better access to clean water and other services, as well as improved environmental conditions. The Organisation for Economic Co-operation and Development (OECD, 2000), for example, suggests that governance-related issues should be addressed at the metropolitan area level in order to maximize the potential of an urban region. Nevertheless, for its implementation, OECD also recognizes serious limitations such as the fragmentation of administrative jurisdictions, strain on the financial and fiscal ability of local authorities, and lack of decision-making processes that are transparent and accountable, not to mention the inability for institutions to achieve coordination and communication.

In order to ensure good water governance, it is necessary to draw lessons by analysing outcomes from cases around the world which can be considered as examples of good or bad governance. In an Asian context of urban water management, water utilities in Tokyo,

Singapore (Tortajada, 2006; Khoo, 2010) and Phnom Penh (Chan, 2009; Biswas & Tortajada, 2010) stand as cases of good governance, where remarkable progress was made within a period of 10 to 20 years, albeit over different periods of time.

What is urgently needed is some 'reverse engineering' to determine why and how these cities have succeeded to make remarkable transformations within limited timeframes as a result of which they have become world class utilities. Similarly, it will be desirable to find out why many other utilities have failed, especially where availability of funds has not been a major issue for the last several decades. Such reverse engineering is likely to throw new light on good water governance practices in the urban sector, which may be very different from the current mainly academic debate on this topic. The Institute of Water Policy of the Lee Kuan Yew School of Public Policy and the Third World Centre for Water Management are now conducting such a joint study on the reverse engineering of Singapore and Phnom Penh utilities. Preliminary results indicate that, at least for these two utilities, the current debate on good water governance may have to be reconsidered, rephrased and restructured. Equally, the Asian Development Bank (ADB) and the Institute of Water Policy of the Lee Kuan Yew School of Public Policy are carrying out studies: to identify and analyse case studies of good practices for urban water management in Asia; to host regional meetings on improving water governance; and to develop an executive leadership programme for urban water managers.

Panel Discussions

During the panel session, it was discussed that there is not 'one' global structure that can be put in place to ensure good governance in the water sector. This is because societal considerations are different in the various countries and thus each requires their own specific solutions. Water governance requires particular policies including those on appropriate pricing for each city. It requires developing effective partnerships between various public entities and between public entities and the private sector. And it requires seeking the engagement of society for support and active involvement. Broadly speaking, water governance should consider both inter-sectoral water issues and cross-sectoral concerns such as those from the energy, agricultural or environmental sectors. In terms of food security, considerations could broaden from a narrow view on only agricultural water use, to policy considerations on its overall impact on water management, including the possible role that virtual water can play on the whole equation of food security.

It was agreed that the implementation of the concept of water governance requires that the entire water cycle be considered as a system. An overall 'system thinking' mentality is urgently needed wherein the entire water cycle is considered and managed under one institution and not fragmented under several institutions, each with different agendas and clients, and overlapping jurisdictions. This often contributes to inter-institutional rivalries and conflicts which, in turn, result in poor governance. Within this 'system thinking', centralized systems could support the development of strong local institutions that are able to provide better service to all sectors of society. This includes the role that central governments can play to ensure that local governments have access to financial markets, and that targeted subsidies are available so that local governments can plan for their operation on a longer term basis.

Regarding partnerships, it was discussed that functional and efficient institutions require partnerships between public and private sectors that combine healthy competition and effective regulations. In fact, regulation and operation of water utilities by the public sector have provided them with the opportunity to develop the expertise necessary to effectively monitor and regulate the performance of the private sector. In terms of human resources, should water supply be controlled entirely by the private sector, governments would lose the opportunity to develop human resources that are skilled in the overall management of water utilities and thus on the efficient provision of services.

Constraints on efficient and functional water utilities include not only lack of experienced personnel and inadequate financing, but the fact that most state actors do not have clear targets in terms of water supply and sanitation, and many times they do not even have a clear appreciation of the resources under their control and thus on the most efficient use of their resources. In terms of pricing policies, there is no agreement as to what is the best alternative. For example, sustainable cost recovery or marginal cost pricing is part of an on-going debate, and no consensus is likely to be reached in the near future. There are differences of opinion on government interventions, subsidies, marginal pricing, water conservation taxes, flat water tariffs, and the ability of the poorest population to pay. These are best decided on a case by case basis, depending on the specific conditions of each city and their social, political, economic and environmental requirements.

It was emphasized that two important pillars of governance—transparency and accountability—should help to build an efficient system that is sustainable over the long term. Instruments such as dissemination of knowledge and information are considered to enhance accountability and transparency of operations both in the public and private sectors and, as such, they should be encouraged both within the institutions and also towards the public. This is because governance-related issues are not the exclusive domain of the government any more; the increasing participation of the private sector and societal actors has made decision making an exercise which has proved to be exceedingly complex and which still needs to be built.

In terms of the provision of services, the arguments were that they include not only operational alternatives, but a very complex series of issues that have to be carefully adapted to each specific case. These include management aspects, governance concerns, and mostly shared decision making between several actors in order to suit local conditions, develop a mix of policies and instruments, and encourage human resource skills where the acceleration of socio-economic development is affected by the quality of human resources available. For example, Singapore, a city state that has managed to close the water cycle most efficiently, has implemented specific water demand and supply policies which would not be easily replicable as a whole elsewhere, but which are worth learning from as an example of the successful implementation of water governance. In an Asian context, the Institute of Water Policy of the Lee Kuan Yew School of Public Policy and the Third World Centre for Water Management have embarked on a study of reverse engineering in order to identify from an implementation viewpoint the essential factors contributing to good governance in the urban settings of the successful Asian cities of Singapore, Tokyo and Phnom Penh. Other non-Asian cities which have good urban water governance systems will also be considered. Instead of focusing on the theory of water governance and/or on defining what constitutes good water governance in an urban context, the study is focusing on the essential qualities that have contributed to good water governance in the above three cities. The results of this study will be an educational experience for other cities which could learn the reasons why these three cities have achieved good governance when most others have failed.

The analysis of Phnom Penh has been finalized and can be seen elsewhere in this issue (Biswas & Tortajada, 2010).

Final Remarks

Governance, and in this specific case water governance, embraces, in addition to rules, regulations and institutions, value-related issues such as responsibility, accountability, transparency, equity and fairness. This adds an enormous complexity to the unsolved challenges that already face the implementation of water policies in an increasingly complex environment.

Since water is an intrinsic part of the overall development process, planning and implementation activities in the water sector are necessarily dependent on other economic and productive sectors where needs and interests of multiple public, private and societal stakeholders intersect with each other. Water management, and now its governance, is thus under multi-dimensional pressure. Governance of the water sector must thus take into consideration the continual changes, both from within and outside the water sector, which affect it and on which water professionals have limited control. The implementation of any long-term vision on water governance will require the understanding of the immense changes and challenges that are likely to be faced in the coming years, and on defining the ways in which these can be best understood and addressed for the overall socio-economic benefits of the countries and their citizens.

Acknowledgements

The notes taken by Mr. Peng Shugang, Mr. Vinesh Kumar and Ms. Chong Su Li are gratefully acknowledged. They were students of the Master degree of Water Policy and Governance course at the Lee Kuan Yew School of Public Policy, Singapore, at the time this workshop was organized.

Notes

1. In terms of definitions, each organization has described 'governance' in terms of its own interests. For example, the Organisation for Economic Co-operation and Development (OECD) considers governance to be the use of political authority and exercise of control in a society in relation to the management of its resources for social and economic development (OECD, 1995). The World Bank defines governance as the way in which power is exercised in the management of the economic and social resources of any country. It takes into consideration the countries' political regimes, the processes by which authority is exercised in the management of economic and social resources for development of the nation, and the capacity of governments to formulate and implement policies and allocate functions. The World Bank has set three goals for good governance which include empowering citizens to hold governments accountable through participation and decentralization; enabling governments to respond to new demands by building capacity; and enforcing compliance with the rule of law and greater transparency (World Bank, 1994).

   The United Nations Development Programme (UNDP) considers governance as the exercise of economic, political and administrative authority to manage the affairs of any country at all levels, and the mechanisms, processes and institutions through which citizens and groups articulate their interests, exercise their legal rights, meet their obligations and mediate their differences (UNDP, 1997). Similar to other international organizations, the UNDP considers that good governance requires participation, transparency and accountability, and that it should promote the rule of law. Governance encompasses the state, but also the private sector and civil society organizations.

   The Commission on Global Governance regards governance as a multi-level phenomenon. It describes it as the sum of the ways in which individuals and institutions, public and private, manage their common affairs through a continuing process that accommodates conflicting and diverse interests while fostering cooperative actions. In other words, governance refers to the many channels through which 'commands' flow in the form of goals framed, directives issued and policies pursued. It includes formal institutions and regimes empowered to enforce compliance, as well as informal arrangements to which people and institutions either have agreed to or perceive to be in their interest (Commission on Global Governance, 1995).

The European Union has established its own concept of governance in the White Paper on European Governance (Commission of the European Communities, 2001). The term 'European Governance' refers to the rules, processes and behaviour that affect the way in which powers regarding openness, participation, accountability, effectiveness and coherence are exercised in the European Community. Multi-level governance identifies the challenge of articulating the action of independent public actors at different geographic levels towards shared objectives. Since the idea of governance highlights the involvement of regional, local and non-governmental actors in the policy-making process, it is increasingly clear that the success of decision making and the acceptability of its rules depend on such actors being involved.

2. This section is based on the presentation of David Molden at the workshop. The final paper on this topic can be seen elsewhere in this issue (Molden *et al.*, 2010).

References

Biswas, A. K. & Tortajada, C. (2010) Water supply of Phnom Penh: an example of good governance, *International Journal of Water Resources Development*, 26(2), pp. 157–172.

Chan, E. S. (2009) Bringing safe water to Phnom Penh's city, *International Journal of Water Resources Development*, 25(4), pp. 597–609.

Commission of the European Communities (2001) *European Governance: A White Paper* COM (2001), 428 final Brussels: Commission of the European Communities.

Commission on Global Governance (1995) *Our Common Neighbourhood* (Oxford: Oxford University Press).

Khoo, T. C. (2009) Singapore water: yesterday, today and tomorrow, in: A. K. Biswas, C. Tortajada & R. Izquierdo (Eds) *Water Management in 2020 and Beyond*, pp. 237–250 (Berlin: Springer).

Kooiman, J. (2003) *Governing as Governance* (London: Sage Publications).

Molden, D., Lautze, J., Shah, T., Bin, D., Giordano, M. & Sanford, L. (2010) Governing to grow enough food without enough water—second best solutions show the way, *International Journal of Water Resources Development*, 26(2), pp. 249–263.

OECD (1995) *Participatory Development and Good Governance* (Paris: OECD).

OECD (2000) *The Reform of Metropolitan Governance* (Paris: OECD).

Rhodes, R. A. W. (1996) The new governance: governing without government, *Political Studies*, 44, pp. 652–667.

Rogers, P. & Hall, A. W. (2003) *Effective Water Governance* TEC Background Papers No. 7, Stockholm: Global Water Partnership.

Tiihonen, S. (2004) *From Governing to Governance: A Process of Change* (Tampere, Finland: Tampere University Press).

Tortajada, C. (2006) Water management in Singapore, *International Journal of Water Resources Development*, 22(2), pp. 227–240.

Tortajada, C. (2007) Water governance in the MENA region: critical issues and the way forward—the case of stakeholder participation and public awareness. Paper presented during the Second Partners' Forum on "Water Governance in the MENA Region: Critical Issues and the Way Forward", organized by InWEnt Capacity Building International, Germany, and the Arab Water Council, Cairo, 23–26 June 2007.

Tortajada, C. (2010) Water governance: a research agenda, *International Journal of Water Resources Development*, 26(2), pp. 309–316.

UNDESA, UNDP & UNECE (2003) Governing water wisely for sustainable development, in: United Nations, *World Water Development Report: Water for People, Water for Life*, pp. 369–384 (Paris: UNESCO).

UNDP (1997) *Governance and Sustainable Human Development* (New York, NY: UNDP).

UNDP (2004) *Water Governance for Poverty Reduction: Key Issues and the UNDP Response to Millennium Development Goals* (New York, NY: UNDP).

World Bank (1994) *Governance: The World Bank's Experience* (Washington, DC: World Bank).

Water Governance: A Research Agenda

CECILIA TORTAJADA
Lee Kuan Yew School of Public Policy, Singapore; International Centre for Water and Environment (CIAMA), Zaragoza

ABSTRACT *This paper summarizes the discussions held during a brainstorming session on water governance organized as part of the First Global Water Policy Dialogue of the Institute of Water Policy of the Lee Kuan Yew School of Public Policy, Singapore, and the Third World Centre for Water Management, Mexico. The objective was to formulate a priority research agenda on water governance which would focus on vital issues ahead such as linkages and drivers of change in the water sector and the type of institutions and instruments that are and will be needed in order to face the increasing challenges in the sector.*

Introduction and Background

Given the interest in the topic of water governance, the complexity of its implementation, and the consideration of the priority issues needed to ensure good water governance, two interrelated events were organized in the Lee Kuan Yew School of Public Policy as part of the 2009 Global Water Policy Dialogue of the Institute of Water Policy. The first of these two events was a workshop within the framework of the Singapore International Water Week which focused on critical issues related to water governance (Tortajada, 2010). The second one was an invitation-only workshop of international experts the objective of which was to develop a priority research agenda that could help the water profession to better understand what constitutes good water governance and how best to implement it.

The participants in the brainstorming session were from different water sectors, countries and disciplines, and from both national and international institutions. They were leading and influential figures from academia, the public and private sectors, and non-governmental organizations. The experts were selected exclusively on the basis of their knowledge in the field of water governance. They were invited in their personal capacities, and thus all the views expressed were their own, and not of their institutions.

The participants were requested to send an outline of their personal views before the session on what they considered to be the three priority issues of good water governance, as well as the reasons thereof. These views were circulated to the entire group in advance to promote a free and spirited discussion.

The brainstorming session started with brief presentations on the critical issues each expert had proposed earlier, justifying the rational behind the selections. Following the presentations, there were in-depth discussions on the overall topic of water governance. Finally, each one of the participants identified only one specific priority area of research on water governance, explaining the reason why he/she considered it to be the most important area from the own perspective.

The discussions were wide-ranging and considered different issues, approaches, successes, failures, constraints, road maps, and research agenda on the overall topic of water governance. In order to focus the discussions further, two questions were subsequently posed to the group: (1) what in their view were the two most important issues related to water governance; and (2) what did they consider to be the two most important topics in terms of research in the area of water governance?

Vital Issues Ahead for Water Governance

The group discussed a diverse series of topics based on their broad knowledge and years of experience. Issues such as policies, institutions, regulations, paradigms and their implementation, development of infrastructure, tariffs and subsidies, use of technology, as well as capacity building were considered to lie at the heart of water governance constraints.

It was recognized that in most of the developing world, policies currently in place have not reached the expected results in terms of water management and governance due to reasons that range from the usual constraints like short-term planning, shortage and inappropriate use of financial resources, to lack of institutional and human capacities. A main lacuna was considered to be the lack of vision to develop strategies that go beyond short-term political and economic considerations, that focus on the long-term needs of the water sector for alleviating poverty, improving quality of life, and protecting the environment. Other reasons identified were legislation and regulations that lack the necessary implementing instruments, inefficiency inherit in public expenditure on water services, subsidy regimes that do not promote innovative ideas, and water organizations that do not attract or retain staff with the skills required for efficient service delivery.

For the countries in the developed world, even though the water sector provides better water supply and sanitation services, management-related issues need serious attention. Some examples include water quality problems, mainly in terms of lack of control of non-point sources of pollution, and the overall inability to coordinate the management of water resources with other resource sectors and development- and environment-related issues.

Concerns that are equally relevant for both developed and developing countries include insufficient investment for infrastructural development and rehabilitation, inadequate coordination and communication between and among agencies, and lack of mechanisms for effective participation of different partners or stakeholders at different levels of decision making. A main problem identified is the lack of realization that water is a cross-sectoral issue, and thus, that its governance increasingly depends on policies in other sectors.

There was a general consensus on the need for critical, objective and in-depth case studies of urban water governance, the reasons for their success, and governance processes used. Only with such an approach will it be possible to identify diverse 'models' of urban water governance that could be considered for possible use in different parts of the world, with appropriate modifications to suit site-specific local conditions.

While the experiences and priorities of the different experts varied, some clear themes emerged as vital areas to focus on regarding water governance both at present and looking towards the future. These are explored below.

Linkages and Drivers of Change in the Water Sector

There are increasing numbers of linkages and drivers of change which are already having an impact on the water sector and its governance. These driving forces include not only population and urbanization, economic growth, energy generation and agricultural production, but also far more complex issues like globalization, free trade, immigration, advances in technology (biotechnology and desalination, for example), changing management paradigms, and evolving social attitudes and perceptions. Water is increasingly becoming a central consideration for food, energy and environmental security and thus is having significant implications for future human development at the global, regional, national and sub-national levels. As such, it is essential that policies and strategies for the water sector are not developed in isolation, as it is mostly done at present, but viewed in a broader and systematic manner so that their impacts on other sectors are considered, and vice-versa.

The pace at which changes in development and technology are happening at present in the world makes it essential that the implementation of changes in water governance can match the constant evolution of the other sectors. However, the required reforms in the governance of the water sector are taking place very slowly, partly due to the lack of administrative capacity, but also partly due to the lack of incentives or interests for actors to change, an issue which may be economic in nature but which also may be related to leadership capabilities. The slow pace of formulating a much needed vision for the future governance and management of water, which can actually be implemented, is now a major constraint.

Within the water sector itself, it is necessary to link efficiently the several diverse components of the water supply, reuse and sanitation loop, with administrative and financial strategies along with a decentralized decision making and implementation at the appropriate regional or local levels. Administrative decentralization can be very effective if successfully implemented along with financial and technical decentralization. Any such process, if it is to be effective, will require a building up of the local capacity. Otherwise, there can only be an extension of bad governance, with the expected poor outcomes.

Importance of Institutions

Properly structured and managed institutions should be able to reach both the macro- and micro-levels of any society in terms of policies and services. Institutions can also assist in improving human welfare through better services, generation of employment, advances in knowledge, education and training, dissemination of information and improved communication. The roles water institutions can play in the future through their interactions with institutions of other sectors and various societal groups, can indicate how much these institutions can contribute to improve the quality of life of the population, promote social cohesion, and protect the environment, through the good governance of water resources.

Functioning institutions can accommodate mixed demands and interests, not only from the public sector, but also from the private sector, academic and research institutions and non-governmental organizations. Institutional arrangements can directly contribute to the success or the failure of the formulation and implementation of good water governance practices, including the design, execution and management of water projects.

In terms of water governance in the urban areas, many failures can be attributed to poor institutional and financial structuring of programmes and projects. Regarding governance of water utilities, there was a general consensus that the structure of utilities could often be of less importance than their ability to meet key objectives such as production and delivery of high quality drinking water, financial viability, and sustainable business models.

In this sense, water tariffs emerged as an important requirement of good and sound water governance. Generally, tariffs cover about 10% of the total operating costs of many water utilities in the developing world. Several arguments were thus made on the need to price water properly:

- If there are no tariff reforms, there is little capital investment in water infrastructure since governments simply do not have adequate or unlimited resources to provide heavily subsidized water.
- Tied to the previous point is the relatively backward nature of technology in water since there is no market incentive for innovation given the low rates of returns.
- From a consumer point of view, if there are no tariffs, there are no incentives for conservation.

Institutions also need to be accountable. Nevertheless, an important issue to discuss is accountable to whom: elected representatives, governmental checks and balances, or the general public? This has implications in terms of modalities and processes, since the different groups will require different types of information for accountability requirements.

It was also noted that there should be clearer implementation and planning guidelines since so far there is very weak articulation of specific policy targets and outcomes with respect to service delivery. Even when these goals are defined, follow up activities to achieve them remain sporadic. Too often, accountability takes the form of building an asset, whereas the important tasks of operation and maintenance as well as acceptable service delivery continue to receive inadequate attention.

Strong institutions which are governed, planned and financed adequately are therefore vital for providing reliable water services, including ensuring access to clean water for the people who may not be able to afford to pay for the full price for the service received. Nonetheless, governments, and mainly politicians, often underestimate the actual cost borne by the people in getting access to water, such as collecting it from wells and rivers, or paying providers at high costs for poor quality water. There is often the erroneous perception among many politicians that there is a heavy political cost for charging the full price for a reliable supply of water when, in fact, this could actually result in increasing public support since many people may have to pay higher economic and social costs by buying poor quality water from private water vendors, the supply of which is also unreliable. Consequently, issues of equity and poverty could be of structural nature. A system with heavy subsidies could thus be changed to one where there is full cost recovery with subsidies targeted only to the poor.

Finally, the study of water institutions needs to be governed by good information. This includes reliable data on water supply-demand relationships, as well as related indicators, which would provide benchmarking for common understandings and inter-utility comparisons. At the same time, there should be objective analysis of current practices in order to learn from them.

Holistic Views and Their Complexity for Implementation

Holistic approaches are considered to be of fundamental importance to natural resource management, including water. However, both developed and developing countries have faced, and continue to face, serious constraints which have made implementation of the current paradigms either impossible or exceedingly complex, including all those related to holistic approaches.

For decades, water management trends have included, at least in theory, the implementation of concepts such as sustainable development, river basin management, integrated water resources management and now lately, water governance. The drawback has been that none of these paradigms have thus far rendered practical results at a perceptible scale. This should encourage water policy-makers, managers and users to rethink how these paradigms could be modified, or even replaced, so that they could be implementable under different contexts. Rather than naively assume that it would be possible to achieve the implementation of any of these paradigms without drastic changes, mainly at the institutional and regulatory levels, it would be more practical if policy-makers, managers and practitioners made an effort to prioritize their requirements, objectively identify the constraints on the implementation of the current paradigms and how to overcome them, and then focus on improving management and governance of water resources in real terms.

So far, inappropriate and inefficient resource management practices often prevail. A long-term futuristic vision for a much needed broader perspective on key-issues related to management of water, and institutional arrangements and coordination between resource policies of various sectors, are missing. Performance indicators are often disregarded, and there is questionable public sector leadership. There is almost no consideration to develop new capabilities and knowledge to identify long-term water governance issues, issues which will certainly be very different from present and past matters.

Additionally, despite the fact that the policies on water, agriculture, energy, industry and environment are closely interlinked with each other, water is managed as if it was independent from all other sectoral issues. In fact, the missing link between the water sector and the other sectors is one of the big gaps in the management and governance of this resource that will have to be addressed sooner than later. Tackling problems, individually or partially (for example without explicit consideration of land use and other related issues) has been the practice for years mainly because of administrative, institutional and legal convenience. But experience has shown that the *status quo* will not yield optimal results in the future. Decision-makers will have to consider before long how to develop a holistic systems approach that improves water governance and which benefits society significantly.

Political will is more than a buzzword. It is an essential requirement to ensure that water resources and urban planning are planned and managed strategically in relation to the development of cities as a whole. This may sometimes mean safeguarding the resource from pressure to develop. For example, in Singapore there has been a determined political will for decades to protect the central catchment area in spite of significant pressure from developers, because development would mean increasing water requirements in the future. Part of good water governance, therefore, has to come from balancing land and water resources in terms of future development plans, an issue which will make the governance of water increasingly difficult and complex in the years to come.

A holistic view of the governance of urban water resources also includes human access to clean and drinkable water from the tap. This issue, as well as the development of strategies to deliver services to every one, especially the poor, has much to gain from an efficient system which includes not only the reliable provision of clean drinking water, but also fair and equitable water pricing for people at all socio-economic levels. What is needed is universal access to clean water at affordable prices for the totality of the population, and it can be achieved with better governance practices.

The holistic view of the governance of water resources should also allow policy-makers to take virtual water into account. Neglecting this component of water could be an economic miscalculation since it may not reflect the true costs (or price) of water. In addition, the concept of the food self-sufficiency of nations should give way to ensuring food security, with consideration of water requirements for the agricultural sector.

The increasing complexities of the problems associated with efficient water resources management and governance, and growing societal interests in water-related issues, have necessitated a broader and systemic approach that cannot be provided by the engineers and/or by the economists alone. Broadening participation to disciplines and stakeholders other than those related to the public sector at the central level may result in the consideration of different alternatives which the present institutions have been unable to frame so far. It is thus crucial to consider a whole range of partnership modalities with industry and business, as well as groups from academia and society, which vary depending on the specific situation. An example is the '3-P model' in Singapore where public, private and 'people' sectors are included in the equation. It is important to keep in mind that transparency and dissemination of information are fundamental requirements to set the foundation of any partnership so that it can render the expected results both in the long and the short terms.

An overall constraint is that people in general are not aware of the critical nature of water issues that the world is facing; many times debates focus on trivial issues. Nevertheless, engagement of local communities and exchange of views and information is fundamental to obtaining public support. In terms of the provision of water services and the governance of supply systems, the challenge for governments is to decide how and when specific responsibilities can be delegated to several stakeholders, ensuring at the same time the efficiency of the systems. This most probably would require clear regulations as well as the introduction of instruments such as incentives and disincentives to encourage responsible behaviour, whether on the part of industry, business, civil groups, or individuals.

The processes to ensure real and efficient participation of different stakeholders and the roles they could play according to their capabilities, interests or relevance, also add to the complexity of good water governance. Stakeholders' participation in the water sector is a topic that has been extensively discussed and promoted at various national and international levels. Irrespective of the rhetoric, however, not very much has been achieved to ensure its real implementation, or even prove definitely that it always improves water governance. A reason for this may be that the participation of stakeholders has become an end in itself in many cases, with the end objective, the achievement of good water management and governance, being overlooked by those involved in decision making.

Institutions at the international, national, regional and local levels have yet to realize that in order to achieve an effective and realistic participation by the several stakeholders, many other needs like education, training, information and communication must be met concurrently. Participation for the sake of participation may produce results which are not even beneficial for good water governance. It is also necessary to develop flexible processes on how best to reconcile the different interests of the stakeholders which often could be polar opposites. Without clear ideas of processes available to mediate such differences of interest, stakeholder participation has the potential to improve the governance system, but also to produce complete paralysis.

Research Agenda on Water Governance

For the discussion on the research agenda, each participant of the workshop was requested to indentify one priority research area on water governance. This was followed by a general discussion on what should be an overall priority agenda on the topic of water governance (names and positions of the participants who took part in this session and the areas recommended, can be found in the extended report prepared for this session (Tortajada, 2009)).

The following topics were considered important as part of a priority research agenda. They are not presented in any specific order or priority because their importance and relevance may vary from one country to another and may even vary over time.

It is vitally important that research conducted provides a fundamental theoretical basis for a conceptual framework of water governance that is implementable. By developing a research agenda on water governance, and with the results obtained, there is the potential to shape how water is governed in the future.

1. Future-oriented governance research. Water governance will have to change radically to manage cost-effectively and in a timely manner the impacts of new drivers such as globalization, free trade, migration problems between countries and within countries and information and communication revolutions. In addition, the nature of the existing drivers will also change significantly in the coming years. Among the changes of existing drivers such as population are the declining population in many Western countries, population stabilization in many developing countries within the next four to five decades, changes in the age structure of the population (for example, by 2030, China will have more elderly people than the entire current population of the United States), and rapid growth of small-to-medium-size cities. A very important topic to study will thus be the direct and indirect water requirements in terms of quantity and quality, and the indirect water-related impacts through sectors like energy, agriculture, environment and health, including how the current structure of water governance should change to identify and respond to these expected changes, and how such changes should be managed within the water sector.
2. Urban water governance. It is essential to identify 6–10 case studies of cities where significant improvements in water governance have taken place over the course of a decade during the past 50 years. These could be analyzed through reverse engineering to determine the key enabling factors that allowed them to make such progress within such a limited period. A very important issue will also

be the replicability potential of these success stories in other cities of the world.

3. Governance of the water sector as a whole. Participants felt that difficult though it has been, governance of urban water systems, compared to overall water resources management and governance as a whole (including water for agriculture, energy, environment and other sectors) is relatively simple since the objectives of urban water management can be clearly defined without any controversy or serious debate. In contrast, overall water management and governance is a complex issue, where objectives are many, some of which may be subjective in nature and even conflict with each other. Thus, while it is relatively simple to identify good and functional governance of urban water systems, it is comparatively difficult to identify what are the best examples of comprehensive water management and governance. However, irrespective of the complexities and the difficulties, an attempt should be made to identify cases where the management of specific systems can be considered some of the best, and the current governance processes can be considered as sustainable over the long-term. These cases should then be studied to determine why and how good water governance has been achieved, what are the individual strengths and weaknesses, and what the potential replicability is in other parts of the world after necessary modifications to suit the local conditions.
4. Institutions. What are the different types of institutional arrangements that may contribute to a functional and coordinated approach to important water-linked sectors like energy, agriculture and environment, including how policies, which are compatible and do not conflict with each other, in the four interlinked sectors can be developed and implemented.
5. Tariffs and subsidies. There is a wide misconception on the impacts of tariffs and subsides, both among politicians and water professionals. Research is therefore needed on how best to structure tariffs, and on identifying the best types of subsidies that simultaneously satisfy the objectives of the financial sustainability of institutions governing water, equity issues, management of demands, and water conservation.

Some of the issues discussed during the brainstorming session are being incorporated by the Institute of Water Policy in its own research agenda in partnership with several other institutions in different parts of the world. By disseminating the ideas discussed in this brainstorming session, it is expected that more institutions will also consider taking them into consideration in their own programmes and projects.

Acknowledgements

The help of Leong Ching Ching, doctoral student at the Institute of Water Policy, Lee Kuan Yew School of Public Policy, Singapore, for the preparation of the Extended Report of this session is gratefully acknowledged.

References

Tortajada, C. (2009) Extended report of the brainstorming session on water governance, Institute of Water Policy, LKYSPP, Singapore.

Tortajada, C. (2010) Water governance: some critical issues, *International Journal of Water Resources Development*, 26(2), pp. 297–307.

Index